West Virginia
1860 Agricultural Census

Volume 4

Transcribed and Compiled by
Linda L. Green

WILLOW BEND BOOKS
2007

WILLOW BEND BOOKS
AN IMPRINT OF HERITAGE BOOKS, INC.

Books, CDs, and more—Worldwide

For our listing of thousands of titles see our website
at
www.HeritageBooks.com

Published 2007 by
HERITAGE BOOKS, INC.
Publishing Division
65 East Main Street
Westminster, Maryland 21157-5026

International Standard Book Number: 978-0-7884-4488-3

Introduction

This census names only the head of the household. Often times when an individual was missed on the regular U. S. Census, they would appear on this agricultural census. So you might try checking this census for your missing relatives. Unfortunately, many of the Agricultural Census records have not survived. But, they do yield unique information about how people lived. There are 48 columns of information. I chose to transcribe only six of the columns. The six are: Name of the Owner, Improved Acreage, Unimproved Acreage, Cash Value of the Farm, Value of Farm Implements and Machinery, and Value of Livestock. Below is a list of other types of information available on this census.

Linda L. Green
217 Sara Sista Circle
Harvest, AL 35749

Other Data Columns

Column/Title

6. Horses
7. Asses and Mules
8. Milch Cows
9. Working Oxen
10. Other Cattle
11. Sheep
12. Swine
14. Wheat, bushels of
15. Rye, bushels of
16. Indian Corn, bushels of
17. Oats, bushels of
18. Rice, lbs of
19. Tobacco, lbs of
20. Ginned cotton, bales of 400 lbs each
21. Wood, lbs of
22. Peas and beans, bushels of
23. Irish potatoes, bushels of
24. Sweet potatoes, bushels of
25. Barley, bushels of
26. Buckwheat, bushels of
27. Value of Orchard products in dollars
28. Wine, gallons of
29. Value of Products of Market Gardens
30. Butter, lbs of
31. Cheese, lbs of
32. Hay, tons of
33. Clover seed, bushels of
34. Other grass seeds, bushels of
35. Hops, lbs of
36. Dew Rotten Hemp, tons of
37. Water Rotted Hemp, tons of
38. Other Prepared Hemp
39. Flax, lbs of
40. Flaxseed, bushels of
41. Silk cocoons, lbs of
42. Maple sugar, lbs of
43. Cane Sugar, hunds of 1,000 lbs
44. Molasses, gallons of
45. Beeswax, lbs of
46. Honey, lbs of
47. Value of Home Made Manufactures
48. Value of Animals Slaughtered

Table of Contents

Putnam County, West Virginia
1860 Agricultural Census

The University of North Carolina at Chapel Hill filmed the 1860 agricultural census for Putnam County from originals at the West Virginia State Archives under a grant from the National Science Foundation in 1963.

Columns 1, 2, 3, 4, 5, and 13 represent the following information on the census:
1. Name of Owner, Agent or Manager of Farm
2. Acres of Improved Land
3. Acres of Unimproved Land
4. Cash Value of the Farm
5. Value of Farming Implements and Machinery
13. Value of Livestock

Pages are out of sequence in this county. They were transcribed in the order in which they appeared on the microfilm.

J. Hendricks, 30, 170, 200, 10, 206
M. A. Somers, 80, 38, 5800, 312, 883
Jno. Dudding, 40, 33, 2000, 75, 311
M. S. Morris, 80, 20, 4800, 200, 490
W. A. Alexander, 200, 130, 18000, 800, 2870
C. J. Bowles, 60, -, -, -, 396
J. Cartmill, 40, 140, 5000, 30, 70
W. Landshaw, 15, -, -, 8, 90
T. Ellis (Ellie), 400, -, -, 140, 1944
A. R. Wissons, 85, 205, 7000, 100, 395
B. E. Dawson, 16, 84, 800, 5, 200
T. Mason, 45, 75, 1000, 15, 150
G. Fredericks, 45, 58, 900, 100, 586
A. G. Ludman, 35, 45, 500, 10, 257
P. Powers, 40, 80, 1200, 10, 205
J. Ripetoe, 100, 456, 3000, 125, 326
W. D. Meeks, 50, 100, 1500, 65, 268
S. Childers, 40, 200, 1000, 20, 223
J. Childers, 14, 212, 800, 5, 144
E. T. Sines, 30, 272, 700, 50, 351
Geo. Hicks, 12, 90, 400, 6, 15
W. McDonald, 25, 75, 1000, 5, 266
W. Gurley, 20, 80, 500, 20, 138
J. M. Deardoff, 30, 44, 700, 10, 358

W. Gurley, 25, 255, 1000, 75, 214
W. J. Marshall, 30, 146, 600, 10, 95
P. Deardoff, 60, 125, 2500, 70, 326
Jas. Guthrie, 13, -, -, 12, 130
H. Kelley, 16, -, -, 8, 100
H. Jenkins, 20, -, -, 10, 140
J. Lighter, 4, 125, 900, -, -
D. Surbaugh, 70, 215, 1700, 40, 200
W. H. Woody, 70, 184, 1500, 150, 328
A. Taylor, 40, 52, 1000, 40, 75
Jno. Smith, 20, 60, 320, 5, 68
J. Hicks, 50, 110, 1000, 30, 276
W. Tucker, 80, 220, 2000, 150, 546
P. Tucker, 10, 120, 400, 30, 200
W. H. Bailey, 50, 50, 1000, 100, 502
G. W. Rask, 20, 195, 800, 16, 304
E. Maddox, 30, 75, 1200, 62, 187
C. H. Deardoff, 16, 132, 1200, -, -
W. H. Maddox, 35, 115, 2000, 10, 232
E. P. Freeman, 60, 140, 2000, 100, 328
J. V. Young, 60, 65, 1000, 30, 317
G. H. Moore, 35, 115, 800, 15, 110
E. Chapman, 90, 43, 1400, -, 45
J. M. Call, 30, -, -, -, 251

P. Minebrenner, 200, -, 2000, 8, 65
Jno. Racer, 25, 750, 2000, 5, 220
W. K. Horton, 40, 71, 600, 40, 156
M. Gorney, 65, 62, 1200, 20, 260
J. H. Payne, 50, 62, 800, 20, 155
A. W. Handley, 30, 370, 2000, 100, 674
W. H. Laine, 8, 192, 300, 8, 100
H. M. Gurley, 25, 75, 1000, 14, 75
A. Stark, 30, 110, 800, 20, 137
Jno. Welsh, 21, -, -, 3, 58
A. Gurley, 35, 65, 700, 20, 90
Jno. Gurley, 20, 220, 1000, 4, 40
W. Gurley, 5, 45, 300, 10, -
J. J. Thompson, 300, 50, 17500, 600, 3425
J. Morgan, 300, 50, 17500, 700, 2875
M. E. Sims, 40, 77, 5850, 50, 213
R. W. Sims, 600, 200, 6000, 300, 1523
R. M. Sims, 50, 200, 5000, -, -
J. A. Blakeney, 50, 200, 1000, 10, 85
M. Whitehell, 50, 200, 1000, 125, 450
H. Buzzard, 60, 1140, 4000, 10, 100
W. B. Mason, 15, 100, 1200, 10, 100
B. Gurley, -, -, -, 10, 90
L. Burdett, 90, -, 4800, 80, 480
W. G. Mintronix(Vintronix), 50, 1450, 4000, 50, 628
P. Gurley, 20, 58, 1000, 5, 275
S. Sims, 15, 40, 500, 5, 26
L. Morris, 15, -, 500, -, 55
M. Supple, 20, 80, 600, 10, 88
Jno. Supple, 40, 47, 600, 10, 115
A. Dudding, 58, 145, 5000, 100, 200
A. Dudding, 25, 55, 860, -, -
R. Johnson, 37, 66, 2000, 80, 236
J. Stewart, 350, 750, 24000, 250, 1500
J. C. Ingham, 30, -, -, 20, 230
B. P. Morris, 80, 80, 10000, 75, 320
C. Meeks, 30, -, -, 10, 242
G. C. Bowzer, 200, 171, 10000, 480, 1415

W. H. Lewis, Sr., 20, 330, 11000, 60, 293
A. L. Melon, 75, 225, 4000, 75, 770
J. S. Morris, 30, -, -, 25, 216
W. Cash, 54, 10, 4000, 160, 235
B. Somers, 25, -, 1000, -, 145
A. Slaughter, 30, 30, 500, 5, 90
E. Narrais (Narrait), 130, 250, 10000, 100, 275
Jesse Gimms (Timms), 30, 30, 1200, 75, 360
L. L. Bronaugh, 100, 130, 8000, 100, 880
W. Frazer, 130, 170, 6000, 100, 1880
S. Frazer Sr., 100, 267, 8500, 100, 848
S. Frazer Jr., 50, -, -, 50, 305
J. Bannister, 8, 52, 120, 10, 82
A. J. Bias, 24, 76, 400, 15, 309
S. Baylis, 15, -, -, 3, 20
B. Bias, 75, 330, 1000, 100, 280
N. Handley, 25, 730, 1500, 70, 405
J. Pretty, 8, 45, 750, 2, 95
T. Handley, 20, 148, 2500, 100, 370
C. S. Poindexter, 200, 100, 3800, 85, 840
Wm. Billups, 100, 40, 2000, 15, 432
T. Billups, 25, 100, 500, 10, 88
J. T. Sonne, 63, 110, 1000, 80, 441
H. Saine (Laine), 14, 400, 800, 10, 218
W. Gibson, 15, 44, 300, 2, 30
E. C. Bailey, 25, 25, 325, 5, 269
J. W. Ford, 35, 200, 800, 6, 109
P. Davis, 30, 130, 600, 6, 175
G. W. Ludman, 30, 170, 800, 10, 155
C. Gibson, 100, 340, 2000, 125, 394
A. C. Taylor, 20, 326, 800, 40, 232
E. Estes, 45, 233, 1000, 10, 325
R. W. Foster, 70, 130, 1500, 15, 810
B. Smith, 60, 50, 2400, 30, 424
_. Hodges, 25, 50, 600, 10, 146
P. Hodges, 25, 166, 600, 12, 125
J. Chapman, 30, 30, 600, 10, 145
W. Poor, 40, 400, 500, 25, 72
D. Jones, 130, 880, 7000, 50, 928

J. Libins, 100, 400, 2500, 50, 771
C. Maguire, 60, 240, 900, 10, 137
K. Maguire, 50, 30, 1500, 100, 322
J. E. Robey, 12, -, 1000, 6, 50
J. W. Bridges, 14,-, 70, 6, 48
Z. Knight, 10, -, 60, 6, 98
M. Gallaspie, 40, 285, 700, 50, 297
Jno. Gallaspie, 25, 55, 800, 50, 133
L. E. Vintronix, 60, 596, 6000, -, -
L. E. Vintronix, 65, 36, 4000, - ,-
L. E. Vintronix, 230, 321, 25000, 850, 2114
A. Frazer, 75, 80, 3875, 75, 941
R. Hall, 260, 80, 11768, 400, 1084
C. Frazer, 100, 50, 4250, 15, 365
E. L. Chapman, 45, 255, 1500, 12, 205
J. M. Dewitt, 30, 120, 1400, 10, 221
Jno. Chapman, 10, -, 200, 5, 70
J. W. Ball, 8, 184, 250, 5, 85
_. Taylor, 8, -, -, 5, 41
R. Forth, 20, -, 100, 20, 146
R. Davis, 10, 150, 500, 12, 132
D. Davis, 16, 134, 300, -, -
N. Deal, 70, 930, 3000, 75, 677
C. Jordon, 30, 194, 1000, 75, 548
J. Taylor, 50, 170, 600, 20, 363
J. Young, 12, 145, 500, 6, 210
J. M. Rucker, 12, 135, 1000, 15, 120
C. Deal, 5, 95, 300, 5, 166
J. R. Milliams, 25, 170, 800, 10, 118
S. Milliams, 12, 359, 700, 2, 120
S. Nottingham, 40, 341, 1800, 100, 412
L. Conrad, Valued, by, Nottingham, 5, 134
J. Forth, 50, 445, 1500, 10, 215
E. W. Reece, 30, 470, 2000, 10, -
W. C. Chapman, 12, -, -, 3, 163
Jas. Gallaspie, 25, 475, 15, 230
S. W. Blair, 80, 469, 1600, -, 25
R. Forth, 50, 350, 2000, 10, 115
E. Smith, 20, 140, 500, 5, 135
C. Quarles, 10, 140, 600, 30, 165
A. Quarles, 60, 739, 2000, 70, 250

A. N. Curry, 150, 380, 1000, 100, 554
R. V. B. Thompson, 70, 150, 3000, 102, 1214
M. Chapman, 80, 30, 1200, 70, 246
Wm. Cyrus, 100, (valued by A. Ellis), 30, 413
J. L. Kinnard, 18, 382, 1000, 6, 30
L. Treadaway, 20, (valued by A. Ellis), 5, 100
T. Erwin, 40, 200, 1000, 30, 327
J. E. Duffer, 5, 95, 300, 5, 67
Jos. Erwin, 15, 85, 300, 10, 187
Jas. Ballard, 250, 250, 500, 20, 200
Calvery McCallister, 150, 126, 4800, 100, 700
Isaac Ellison, 70, 70, 4000, 50, 245
Thos. McCallister, 150, 1000, 5000, 100, 600
Jas. Olliver, 75, 75, 2000, 10, 170
S. W. Duke, 40, 237, 600, 15, 136
J. Duke, 20, 50, 500, 6, 125
J. S. Kirtley, 80, 510, 1600, 100, 366
C. J. Riddle, 50, 280, 1000, 10, 271
E. Harborn, 45, 20, 2000, 150, 334
David Gore, 200, 150, 4000, -, 50
Lewis Johnson, 10, -, -, -, 50
J. D. Milliams, 250, 200, 5000, 10, 98
S. McKinney, 25, 75, 1100, 15, 210
Brice Paul, 40, 70, 750, 10, 230
A. Ellis, 400, 200, 7500, 400, 2675
Geo. Ellis, 21, 100, 800, -, 65
Robt. Martin, 50, 94, 1680, 25, 346
A. Ellis, 25, 50, 750, -, -
Jno. Griffith, 36, 156, 1000, -, 115
Jas. Mithell (Mitchell), 60, 240, 1500, 75, 650
Wm. M. Ellis, 35, 163, 1200, 40, 361
Wm. N. Griffiths, 30, 70, 50, 5, 205
D. Mynes, 70, 38, 800, 50, 200
B. Johnson, 30, 20, 300, 70, 150
C. Allen, 30, 26, 200, 5, 113
A. Ellis, 25, 100, 500, 7, 144
J. A. Henderson, 60, 80, 1000, 60, 165

H. Ellis, 50, 30, 1000, 60, 370
G. W. Mynes, 75, 140, 1000, 100, 450
J. W. Mynes, 45, 93, 1000, 140, 245
A. Bias, 30, 100, 1000, 10, 90
S. L. Billups, 18, 100, 500, 3, 175
D. Libins, 30, 400, 2000, 8, 180
D. P. Savine (Laine), 25, 275, 1200, 5, 110
J. Roberts, 100, 100, 2000, 100, 432
Wm. G. Cyrus, 60, 60, 1000, 5, 150
T. H. Mynes, 80, 800, 1500, 75, 333
A. A. Cyrus, 25, 75, 700, 25, 80
T. Pane, 50, 70, 800, 60, 133
S. Hodges, 100, 96, 2000, 75, 461
J. Young, 50, 150, 1400, 50, 242
A. A. Young, 60, 40, 1200, 90, 496
Wm. J. Mynes, 10, 70, 300, 5, 188
W. R. King, 50, 170, 500, 20, 130
J. Mynes, 20, 100, 800, 10, 160
M. Pursinger, 15, 125, 700, 6, 177
T. Beckett, 16, 184, 500, 5, 50
J. R. Short, 10, 60, 350, 6, 50
J. C. Henson, 40, 15, 500, 100, 325
S. Cyrus, 25, 35, 500, 5, 192
L. H. Ellis, 25, 70, 500, 5, 81
P. Burdett, 30, 50, 400, 8, 200
J. R. Hill, 150, 175, 3000, 25, 1130
O. Briscoe, 35, 115, 800, 7, 39
J. W. Young, 60, 230, 200, 50, 440
J. Flesher, 6, 124, 133, 75, 223
H. F. Flesher, 25, 95, 260, 5, 125
A. J. Allen, 80, 75, 1000, 130, 547
R. H. Roberts, 20, 114, 600, 10, 192
H. Paul, 50, 80, 681, 100, 245
H. Conner, 5, 145, 500, -, -
J. Manning, 15, 46, 300, 6, 105
H. Arbaugh, 10, 94, 500, 10, 20
J. Pane, 12, 78, 500, 40, 160
A. Carpenter, 22, 100, 1000, 5, 84
E. S. Kelley, 16, 46, 150, 5, 145
W. S. Smith, 50, 125, 2000, 5, 15
M. F. Carpenter, 38, 73, 1000, 75, 276
E. Elkins, 60, 240, 3000, 40, 453
L. Stickler, 30, 50, 400, 15, 295

A. Burdett, 125, 275, 4000, 81, 1000
J. McCallister, 40, 224, 2000, 10, 300
R. Grass, 4, 112, 150, 5, 10
E. Grass, 40, 104, 2000, 75, 275
W. Grass, 5, 95, 300, 5, 27
T. Hawkins, 7, 13, 150, 5, 120
A. J. Nicholas, 50, 125, 1000, 100, 493
S. Alford, 40, 175, 1000, 5, 149
C. McCallister, 20, 180, 100, 5, 141
J. McCallister, 80, 250, 1000, 100, 587
E. Wheeler, 20, 20, 300, -, 222
P. McCallister, 30, 70, 800, 10, 455
J. Burnsides, 30, 220, 1000, 40, 170
A. J. Hazzlett, 40, 175, 1000, 10, 194
W. J.Hazlett, 25, 25, 500, 10, 170
H. Cyrus, 25, 35, 360, 66, 140
M. Wait, 45, 15, 550, 5, 380
J. Cyrus, 35, 125, 800, 100, 193
R. Burnsides, 16, 50, 150, 5, 100
E. Polly, 30, 270, 600, 10, 156
P. Billups, 40, 160, 901, 50, 316
J. M. Jordon, 10, 145, 300, 8, 125
Jno. Bias, 27, 40, 500, 3, 85
C. Arbaugh, 30, 210, 1000, 10, 191
Jas. Dodd, 20, 256, 1000, 5, 30
L. B. Wheeler, 8, 40, 110, 2, 45
J. Paul, 40, 90, 1000, 50, 461
A. B. Wheeler, 40, 60, 1000, 10, 230
J. Grass, 27, 113, 700, 7, 146
R. A. Smith, 20, 360, 1000, 4, 70
A. Wheeler, 30, 47, 400, -, -
A. Wheeler, 70, 215, 1500, 75, 414
C. M. Swain, 45, 200, 1300, 100, 200
J. Wheeler, 15, 70, 400, 8, 109
J. Wheeler, 60, 30, 2000, 100, 291
J. Stickler, 100, 600, 3000, 100, 688
J. Henderson, 35, 180, 800, 101, 423
S. Burns, 25, 154, 600, 5, 96
A. Hicks, 12, 88, 500, 5, 27
N. Gannon, 65, 145, 2000, 100, 263
R. J. Smith, 80, 320, 2000, 60, 250

A. Lakeman (Labeman), 15, 225, 600, 3, 173
J. A. Hensley, 35, 45, 500, 5, 250
Jno. Hensley, 60, 245, 1000, 40, 125
Wm. Mathews, 70, 170, 1200, 75, 410
P. Fizer, 30, 260, 1500, 10, 231
Jos. Paul, 40, 41, 1000, 10, 199
H. Miller, 45, 655, 2000, 4, 181
C. Conner, 140, 360, 4000, 517, 1130
J. M. Saine, 12, 188, 600, 2, 30
W. Chapman, 80, 37, 1500, 10, 305
Jno. Burton, 30, 25, 1000, 40, 197
Jno. McCallister, 200, 222, 5000, 60, 520
F. L. Suttle, 10, 172, 400, 6, 95
S. Herran, 50, 50, 800, 5, 53
T. West, 350, 154, 5000, 230, 1389
W. Davidson, 8, 41, 1500, 5, 75
Wm. Love, -, 600, 1000, 500, 1787
A. J. Conner, 150, 100, 2500, 125, 758
Wm. Erwin, 15, 20, 200, 5, 144
A. Erwin, 24, 50, 350, 5, 147
A. Hicks, 30, 470, 1000, 75, 400
M. Jinkins, 5, valued, above, 2, 135
T. Call, 15, 125, 500, 4, 76
J. Mingo, 15, 195, 800, 20, 160
J. R. Middleton, 160, 410, 2000, 195, 727
Jos. Gerry (Terry), 80, 325, 2000, 75, 492
K. J. Handley, 70, 236, 1500, 100, 283
L. Dixon, 30, 147, 1000, 40, 120
W. Hughes, 8, 195, 1000, 20, 305
M. Taylor, 30, 298, 700, -, 10
A. Grant, 7, 193, 300, 10, 58
A. Wordon, 48, 189, 1000, 30, 239
J. Erskine, 100, 350, 3000, 40, 815
W. Handley, 25, 189, 1000, 40, 311
P. Bowyer, 60, 130, 1000, 10, 212
Jno. Seashoals, 173, 75, 5000, 225, 1133
W. Best, 25, 775, 9735, 5, 150

J. A. Best, 40, 160, 800, 6, 50
P. McGitrie, 40, 280, 700, 25, 402
W. Henson, 350, 275, 7000, 185, 1140
C. Handley, 230, 50, 4000, 250, 554
S. Roberts, 100, 240, 1000, 60, 205
A. Hill, 60, 200, 800, 75, 315
R. H. Chambers, 40, 14, 3500, 80, 182
S. Bailey, 25, 200, 350, 25, 196
T. D. Pittsford, 15, 300, 300, 3, 60
H. Hedrick, 15, 300, 300, 8, 225
W. N. Love, 10, 186, 550, 10, 203
W. Fisher, 100, 500, 1000, 50, 463
W. M. Clendenning, 60, 190, 301, 60, 876
J. Painter, 30, 150, 500, 10, 180
T. H. Markham, 35, 75, 1200, 10, 130
J. C. Thomas, 30, 200, 600, 20, 203
T. Markham, 40, 228, 1500, 160, 355
W. Smith, 70, 265, 1000, 10, 366
Wm. Legg, 20, 20, 100, 50, 400
G. E. Allen, 90, 100, 6500, 200, 1185
R. Snell, 375, 400, 35000, 450, 3050
T. B. Higginbottom, 40, 45, 1000, 40, 341
J. Earley, 500, 900, 35000, 600, 1605
F. L. Wuntendon, 20, 120, 1000, 25, 60
L. Vogt, 25, 58, 1400, 20, 143
Jno. Craig, 90, 100, 5000, 20, 255
P. Welsh, 30, 30, 1500, 15, 160
G. W. Hedrick, 40, 50, 3000, 100, 430
R. T. Harvey, 250, 230, 2500, 250, 978
J. C. Harrison, 130, 470, 7000, 200, 570
B. H. Stearett, 150, 156, 15000, 290, 876
J. M. Nash, 220, 113, 12000, 273, 1331
J. Dunfield, 15, 75, 500, 25, 38

5

P. Olddecker, 85, 104, 2000, 100, 456

B. J. Olddecker, 13, -, -, 8, 143

F. E. Custer, 45, 100, 1500, 20, 328

J. Koontz, 48, 142, 1500, 240, 278

J. Garner, 100, 165, 3000, 75, 520

Jno. Koontz, 80, 130, 1500, 50, 120

W. Olddecker, 20, 40, 500, 15, 75

J. P. Henson, 5, -, -, 5, 28

J. Knapp, 25, 25, 500, 15, 130

W. Wallace, 40, 61, 520, -, 27

A. D. Wallace, 6, -, -, 50, 154

Jno. Little, 10, 27, 150, -, 28

W. G. Hoffman, 20, 20, 300, 10, 23

A. Falon, 10, 590, 1200, -, -

N. Hoffman, 30, 98, 820, 60, 190

N. Hoffman, 12, 88, 400, -, -

W. Whittington, 30, 70, 800, 20, 190

S. Whittington, 5, 95, 200, 20, 95

J. D. Higginbottom, 40, -, 400, 35, 195

L. Craig, 40, 500, 2500, 100, 205

P. J. White, 75, 150, 8000, 75, 425

W. Giles, 50, 100, 800, 15, 130

C. Lanham, 40, 70, 3500, 100, 365

A. Librall, 250, 100, 25000, 500, 2152

S. McCoy, 40, 60, 1000, 100, 368

B. Barnet, 30, 270, 1000, 100, 448

W. C. Brown, 110, 250, 8000, 100, 610

J. B. Parker, 65, 335, 5000, 100, 725

J. M. Wallace, 15, -, 100, 35, 105

H. P. Schmitt, 48, 92, 1200, -, 222

F. Schmitter, 41, 30, 1000, 30, 138

J. Carson, 20, 105, 800, 75, 300

G. W. Garner, 20, 80, 200, 5, 45

J. L. Fisher, 65, 35, 1200, 40, 418

P. Pendergast, 12, -, 100, 10, 60

W. Birch, 50, 27, 1000, 50, 383

L. Peirce, 9, -, -, -, 38

R. Cartwright, 12, 23, 500, 5, 35

C. W. Nalle, 10, 190, 400, 5, 82

Jno. Jeffers, -, -, -, 5, 175

J. McChesney, 30, 55, 1000, 10, 328

Jas. P. Burton, 25, -, 100, 5, 140

Jas. Pogue, 12, -, 200, 6, 50

S. Hartley, 80, 270, 1500, 15, 185

S. D. Workman, 16, 134, 675, 20, 336

S. Henson, 20, 130, 1200, 50, 180

J. Garner, 75, 75, 2000, 50, 440

W. Duncan, 20, 210, 230, 40, 50

J. W. Duncan, 4, -, -, 5, 100

Jas. Magraw, 12, -, 150, 5, 85

J. M. Tucker, 20, -, -, 5, 50

A. Rush, 60, 90, 2000, 30, 380

G. Tucker, 40, 153, 1000, 20, 120

Jno. Williams (Milliams), 120, 280, 3000, 100, 513

T. Jas. Giordend, 30, 120, 520, 20, 430

G. Priddy, 20, 510, 1000, 15, 356

E. Jeffers, 35, 65, 1100, 25, 110

J. E. Thomas, 50, 100, 2000, 100, 318

H. P. Higginbottom, 25, 539, 1000, 5, 188

D. Harris, 10, 90, 300, 5, 45

L. Thornton, 25, -, 200, 10, 45

E. Buckalew, 20, -, -, 10, 185

A. Russell, 50, 500, 1000, 10, 135

W. Harrison, 16, 144, 300, 10, 140

W. Hill Jr., 75, 349, 1000, 100, 700

N. Castro, 80, 120, 1000, 120, 405

E. Cutler, 30, 270, 200, 5, 115

M. M. Reed, 15, 185, 500, 30, 88

W. Fisher, 30, 45, 300, 10, 200

H. Wright, 15, -, 100, -, 100

L. C. Fisher, 50, 177, 700, 25, 527

J. Harrison, 30, 220, 500, 5, 70

J. W. Saunders, 25, -, -, 40, 27

W. Saunders, 20, 550, 900, 45, 137

R. Smith, 30, 70, 500, 10, 160

E. Harris, 40, 100, 800, 15, 450

Jno. . Smith, 35, 325, 1000, 12, 150

R. Coleman, 15, 315, 1500, 5, 160

W. B. Sigmund, 20, 560, 1000, 15, 262

W. Kelly, 25, 340, 360, 10, 264

D. Sullivan, 25, 235, 1000, 25, 210

Jno. Hedrick, 25, 262, 300, 25, 261

P. Hedrick, 20, 460, 480, 5, 117
J. Hedrick, 45, 38, 1300, 10, 292
J. Kesser, 15, 88, 300, 5, 50
B. Rayburn, 15, 177, 600, 5, 90
John Blake, 15, 35, 800, 10, 114
J. Steele, 10, -, 150, 10, 25
A. Thornton, -, -, -, -, 300
J. Wade, 20, -, -, 15, 130
D. T. Blake, 10, -, -, 10, 65
A. Atkinson, 150, 237, 25000, 730, 1770
J. Carr, 9, 20, 800, 10, 75
F. Reynolds, 20, 80, 4000, 10, 140
J. W. Wiatt, 130, 220, 10000, 150, 750
F. B. Woody, 20, 80, 3000, 100, 847
R. Long, 40, 210, 8000, 15, 200
F. Ruffner, 600, 1400, 5000, 932, 3535
M. McCowan, 30, 110, 600, 25, 75
A. Chapman, 66, 69, 1000, 80, 419
S. Magraw, 15, 185, 300, 50, 205
Jas. Hutton, 50, 80, 500, 50, 280
Jno. Mayse, 75, 400, 4000, 15, 195
W. A. Beauford, 55, 70, 2200, 20, 388
R. Johnson, 25, -, 1500, 10, 70
T. M. Turner, 5, -, -, 8, 30
J. J. Walker, 15, -, 200, 20, 150
W. Jones, 128, 532, 10000, 200, 500
M. Barbour, 25, 100, 500, 10, 55
E. Cook, 75, 225, 6000, 100, 268
W. Beauford, 120, 173, 6000, 125, 592
A. H. Blake, 51, 68, 3000, 50, 380
J. B. Fanner (Tanner), 60, 160, 1200, 10, 169
Z. Gardener, 20, 60, 500, 15, 118
S. Harmer, 70, 170, 1000, 60, 282
G. Goove (Goone), 40, 330, 1500, 15, 238
Jas. Arthur, 20, -, -, 29, 155
An Cox, 75, 275, 2000, -, -
S. Hereford, 30, 100, 4000, 50, 225
J. Blaine, 70, 730, 15000, 272, 1500
M. R. Harman, 11, -, 580, -, 22

J. A. Harman, 75, 225, 5000, 70, 545
T. Harman, 18, valued, above, 20, 322
Jas. M. Gray, 160, 20, 15000, 500, 1515
R. R. Kramer, 19, -, -, 75, 537
Jos. Bryant, 15, -, 50, 30, 105
J. A. Pursinger, 15, 185, 400, 15, 100
J. Hedrick, 25, -, 100, 50, 288
E. Harman, 15, 85, 125, 8, 60
P. Harman, 14, -, -, 4, 162
A. Rholan, 40, -, -, 130, 410
H. W. Hedrick, 12, -, -, 5, 70
J. M. Caruthers, 30, -, -, 10, -
H. T. Caruthers, 30, -, -, 10, -
Jno. Harman, 20, -, -, 50, 150
J. A. Johnson, 30, -, -, 10, 125
W. H. Thomas, 75, -, -, 35, 270
R. King, 14,-, -, 10, 120
P. Null, 66, 150, 2500, 100, 438
J. W. Lyons, 40, 160, 3000, 120, 300
J. T. Lett, 15, -, 750, 12, 105
A. Asbury, 30, 70, 800, 20, 250
T. H. Carter, 20, -, 1000, 50, 92
M. Higginbottom, 18, -, 900, 30, 345
Wm. Carter, 8,-, 400, 10, 105
J. Higginbottom, 40, -, 2000, 10, 210
Wm. H. Smith, 16,-, 800, 30, 140
J. Higginbottom, 25, 175, 400, 15, 130
J. Hutton, 30, 70, 500, 60, 350
Jas. Higginbottom, 10, 190, 600, 10, 186
A. Ray, 10, 140, 450, 5, 160
J. Crago, 15, 85, 300, 25, 100
W. T. Wilkinson, 15, 85, 300, 20, 190
J. S. Landiss, 15, 215, 50, 10, 110
W. Ray, 25, -, 150, 45, 258
E. M. Ray, 4, -, -, 5, 20
J. Crago, 25, 75, 300, 5, 190
J. Casey, 8, 92, 300, 40, 181
M. & H. Crago, 30, 170, 1000, 40, 238
W. King, 15, -, -, 5, 250
W. Crago, 8, -, -, -, 78

P. Copeland, 15, 35, 150, 30, 273
O. Clarke, 10, -, -, -, 90
Geo. Bailey, 15, 195, 800, 15, 285
G. W. Goff, 20, 80, 400, 25, 140
L. J. Pursinger, 15, -, -, 10, 50
G. J. Landiss, 30, 100, 400, 30, 240
T. Landiss, 20, -, -, 30, 157
T. J. P. Lett, 20, 130, 800, 115, 350
W. Hensley, 25, 75, 1000, 10, 235
B. Hensley, 60, 540, 6000, 50, 169
W. Hensley, 30, 270, 3000, 20, 185
H. C. Carter, 25, 375, 520, 40, 145
Jas. Martin, 60, 40, 700, 70, 336
T. J. Lanham, 20, -, 600, 30, 140
A. B. Harman, 12, -, -, -, 10
Geo. Harman, 55, 140, 4000, 40, 334
J. J. Hays, 5, -, -, 5, 60
L. L. Bowling, 900, 1100, 60600, 775, 4830
N. Amoss, 6, -, -, -, 70
A. R. Rust, 160, 309, 5000, 150, 786
W. Wiley, 30, 95, 1600, 40, 190
R. Simons, 80, 128, 2496, 20, 275
J. Caruthers, 50, 250, 3000, 80, 415
J. Volden, 65, 142, 4000, 100, 480
J. Martin, 5, 25, 200, 25, 51
A. Staley, 15, 30, 400, 10, 210
W. A. Melton, 70, 1030, 4000, 50, 349
W. Martin, 60, 90, 830, 130, 575
E. M. Melton, 100, 350, 5000, 226, 698
E. M. Melton, 60, 23, 1600, -, -
R. Lanham, 40, -, 200, 295, 555

J. Lanham, 10, 20, 600, 20, 202
M. Bailey, 10, -, -, 4, 90
J. D. McCormick, 20, -, -, 6, 169
P. Lanham, 50, 230, 4000, 40, 686
J. Withrow, 25, 24, 300, 5, 42
J. Bailey, 75, 336, 1500, 75, 286
W. H. Lanham, 20, -, -, 5, 68
J. Lanham, 150, 387, 3000, 120, 164
W. Bailey, 40, 62, 1000, 18, 277
M. Milby, 15, -, -, -, 100
W. Thomas, 40, 117, 1000, 35, 110
J. McCormick, 40, 80, 1000, 20, 160
G. Thomas, 30, 127, 1000, 15, 182
L. Goff, 60, 200, 3000, 300, 520
M. D. Truett, 30, -, -, 5, 155
H. Asbury, 16, -, 800, 5, 80
Jesse Thomas, 80, 100, 3000, 20, 100
E. Melton, 40, 590, 2000, 100, 544
W. Melton, 15, valued, above, 5, 190
J. M. Wadkins, 125, 235, 3000, 15, 230
A. Withrow, 25, 71, 1000, 10, 180
A. Baily, 10, -, -, 5, 130
G. McCormick, 18, -, -, 5, 287
S. Baily, 40, 35, 1600, 30, 246
S. Cole, 40, 19, 1600, 100, 193
A. Null, 15, -, -, 10, 295
S. Asbury, 20, 20, 400, 10, 393
J. McLaughlin, 12, 18, 400, 70, 240
E. A. Thomas, 50, 85, 2000, 30, 200
W. D. Short, 75, 400, 6000, 50, 745
Thos. Summers, 80, 20, 5000, 50, 268

Raleigh County, West Virginia
1860 Agricultural Census

The University of North Carolina at Chapel Hill filmed the 1860 agricultural census for Raleigh County from originals at the West Virginia State Archives under a grant from the National Science Foundation in 1963.

Columns 1, 2, 3, 4, 5, and 13 represent the following information on the census:
1. Name of Owner, Agent or Manager of Farm
2. Acres of Improved Land
3. Acres of Unimproved Land
4. Cash Value of the Farm
5. Value of Farming Implements and Machinery
13. Value of Livestock

This county does have some renters.

Josiah W. Teel, 12, 138, 700, -, 97
Samuel Keffer, 23, 21, 1000, 150, 125
Daniel Shumate, 300, 2000, 14000, 200, 1750
Alfred Beckly, 120, 3900, 25000, 100, 258
Johnson Crawford, Renter, Renter, 400, 10, 300
Booker Baily, 50, 150, 1500, 5, 200
John Cook, 100, 352, 2500, 75, 332
William Prince, 130, 100, 3000, 80, 520
William Davis, 30, 127, 1200, 12, 215
Jonathan Howing (Howery), 70, 149, 1500, 55, 456
George W. Baley, 50, 200, 2500, 150, 269
John W. Warden, 20, 270, 1500, 35, 250
Jno. Idens, 25, 75, 500, 35, 176
John P. Hurt, 20, -, 200, 10, 238
John Hurtale, 40, -, 400, 10, 189
Alfred Hurt, 30, -, 300, 5, 55
Andrew Biggs, 20, -, 200, 64, 137
George W. Kincade, 20, -, 200, 3, 243
David Shepard, 20, -, 200, 10, 150

Martin Rogers, 35, 165, 1400, 15, 125
Daniel Williams, 30, 135, 1500, 20, 174
Eli Williams, 20, 98, 1000, 10, 176
Samuel Francis, 100, 999, 6594, 40, 450
Henry Plumly, 45, -, 450, 5, 45
James Scott, 90, 456, 5460, 109, 510
Bartly Pack, 100, 650, 2000, 161, 716
Robert Warden, 20, 510, 800, 15, 235
William H. Carper, 20, 80, 800, 15, 221
George W. Carper, 30, 70, 800, 20, 100
Joseph Carper, 80, 770, 5000, 75, 407
Andrew J. Carper, 30, 200, 100, 20, 200
James H. Redden, 25, -, 200, 10, 300
John Redden Sr., 40, 60, 1200, 5, 348
William Bragg, 75, -, 600, 3, 200
Mathias Redden, 12, 88, 500, 10, 225
William Kidwell, 50, 50, 1000, 10, 175

Charles Furrow, 25, 175, 600, 15, 175

Alexander Waddle, 60, 60, 600, 30, 163

Lewis Meadows, 20, 135, 500, 10, 113

Clarkson Prince, 60, 240, 4000, 65, 200

Lewis Hull, 150, 151, 4000, 100, 335

Timothy Fitzpatrick, 30, 90, 580, 20, 215

Conrad Riffe, 100, 142, 2000, 150, 150

Lewis Graham, 20, 90, 600, 60, 48

James Coffee, 70, 266, 2500, 25, 438

Patric Snuffer 50, 364, 2000, 50, 268

Thomas Pitman, 40, 125, 1200, 15, 30

John Williams, 125, 504, 3000, 150, 416

James M. Bailey, 25, 75, 400, 5, 260

John Evans, 45, 350, 1714, 15, 254

Richard Maynor, 50, 200, 1000, 10, 329

Irvin Stover, 70, 170, 1500, 150, 845

Lewis Williams, 50, 58, 1500, 140, 562

Abram Stover, 45, 172, 800, 25, 138

John Strader, 60, 60, 1500, 35, 201

Joseph Caldwell, 70, 176, 2000, 150, 300

Causdy Smith, 175, 1211, 3000, 75, 531

Joseph Smith, 60, 140, 1500, 80, 228

James H. Spencer, 40, Renter, 400, 5, 165

Jacob Harper, 230, 2250, 8000, 100, 763

William Potsy, 40, 80, 700, 100, 1046

James Potsy, 65, 15, 800, 60, 410

James Potsy Jr., 20, Renter, 200, 5, 135

John P. Jarrel, 35, Renter, 350, 45, 350

Jacob Potsy, 150, 250, 3000, 200, 992

John Bradley, 25, 50, 200, 6, 139

George Bradley, 30, 45, 1000, 6, 170

Nancy Canterberry, 50, -, 300, 12, 284

Meredith Wells, 65, 3935, 4000, 75, 200

Chapman Thompson, 30, -, 500, 12, 500

Charles Lewis, 10, Squaters, 100, 5, -

G. & G. W. Calaway, 135, 1900, 500, 260, 1478

Alexander Bryson, 60, 940, 10000, 100, 670

Lewis McDonold, 350, 1990, 10000, 200, 1577

William T. Vass, 20, Renter, 200, 15, 145

Mary Henderson, 50, 150, 2400, 150, 589

Tilison Shumate, 30, Renter, 300, 40, 300

William Furguson, 280, 470, 15000, 515, 1529

Russel G. Trump, 150, 950, 20000, 125, 1065

Nathaniel McMillion, 30, 97, 1000, 20, 150

Martin C. Mankin, 55, Renter, 550, 6, 150

Owen Snuffer, 170, 380, 3090, 100, 684

Idrogh Davis, 200, 600, 12000, 75, 1077

John Farmer, 125, 275, 4400, 72, 651

Jesse W. Mankin, 100, 900, 10000, 150, 1059

Boswell Vass, 75, 40, 1500, 150, 601

William Oneal, 70, 150, 3000, 145, 827

Newton Shumate, 88, 300, 3000, 145, 802

Charles Hutchison, 47, 823, 200, 5, 137

John W. Gray, 75, 330, 6000, 120, 473

Champ Lester, 70, 600, 2000, 75, 355

Wilson Abbot, 70, 2430, 3500, 80, 1451

Anderson Williams, 15, 103, 200, 5, 69

Abraham McMillin, 27, 163, 400, 59, 192

William Acord, 8, Renter, 80, 4, 35

John B. Turner, 40, Renter, 400, 15, 199

William Daniel, 40, 127, 1000, 20, 217

James Mankin, 50, 100, 200, 200, 221

Orville Wood, 35, Renter, 523, 90, 244

Lewis Cook, 40, 495, 3000, 50, 489

John Fleshman, 30, 200, 1500, 23, 153

Henry B. Clay, 75, 125, 600, 7, 40

John Combs, 200, 800, 5300, 20, 599

Lucy A. Combs, 30, 100, 400, 5, 100

Jeremiah Combs, 12, 58, 250, 5, 42

Jessee Daniel, 40, 460, 3040, 95, 373

William Wills, 25, 100, 300, 5, 70

George Hood, 20, Renter, 20, 5, 100

Hiram Wills, 50, 200, 1000, 20, 340

John T. Seratt, 80, 300, 3000, 75, 582

John H. Thompson, 50, 2000, 1600, 12, 238

William Massy, 35, 200, 1000, 3, 125

John B. Scarborough, 40, 600, 7000, -, -

Alexander Cantly, 25, 75, 300, 40, 161

James Jarrel, 65, 960, 2000, 5,551

William Hunter, 35, 7, 500, 35, 414

John Dickens, 30, 20, 500, -, -

Pemberton Cook, 20, 175, 35, -, -

Andrew Workman, 60, 689, 1500, 50, 250

Henderson Shumate, 50, Rented, 500, 25, 529

Samuel McGinnis, 37, 223, 1000, 30, 202

Albert Jarrel, 20, Renter, 200, 6, 140

Samuel H. Higginbotham, 70, 5000, 6000, 50, 594

John Higginbotham, Renter, Renter, -, -, 400

David Holston, 20, Renter, 200, 10, 183

Stephen Holston, 8, Renter, 80, 5, 15

James Toney, 70, -, 1000, 30, -

Squire Thompson, 20, -, 200, 10, 120

William T. Vandell, 35, Renter, 350, 7, 142

Harrison Jarrel, 14, 26, 200, 35, 129

Henry Williams, 35, 90, 400, 10, 222

Isaac Riston, 60, 400, 1000, 100, 1243

Zachariah Riston, 25, 175, 300, 10, 264

Henry & Charles Grass, 50, Renter, 500, 10, 230

Lemuel Jarrel, 12, 97, 400, 5, 126

Lewis Williams, 35, 265, 1500, 10, 478

Thomas Maynor, 20, 70, 400, 5, 145

Joseph Maynor, 10, Renter, 50, 5, 108

Sampson Stover, 20, 180, 400, 5, 130

Lewis Stover Sr., 70, 550, 800, 10, 301

John Stover, 50, 750, 1100, 10, 254

Burrel Stover, 50, 450, 900, 10, 388

Silas Stover, 16, -, 100, 5, 205

Lewis Stover Jr., 7, -, 35, 2, 118

Daniel & Stephen Stover, 75, -, 750, 60, 514

Arthur Richmond, 70, 1000, 1000, 40, 309

Daniel Bragg, 40, 70, 3000, 30, 160

Abraham Bragg, 170, 730, 5000, 75, 390

John Plumly, 50, 99, 1000, 15, 279

William Plumly, 50, 200, 1500, 50, 276

Washington Plumly, 150, 750, 1800, 20, 576

Jacob Bennett, 100, 2000, 1500, 10, 506

Adam Bragg, 130, 360, 2500, 20, 554

Allen Garten, 100, 460, 1200, 15, 220

Jackson Bragg, 45, 175, 900, 7, 149

Morris Sullivan, 100, 735, 600, 3, 248

John Qunlin, 25, 140, 320, 3, 97

Samuel Richmond, 100, 3992, 2000, 100, 687

Thomas Bragg, 75, 896, 2000, 50, 620

Abraham Meadows, 75, 835, 1500, 20, 353

William C. Richmond, 100, 410, 7000, 125, 272

Hugh Richmond, 40, 872, 1268, 25, 684

Andrew Richmond, 100, 500, 3600, 35, 444

Andrew J. Richmond, 40, 600, 1500, 40, 300

James Richmond, 40, 80, 1500, 40, 355

Moses Richmond, 30, 195, 1800, 10, 330

William Thomason, 98, 204, 3000, 150, 178

Thomas Ward, 40, 360, 1000, 10, 118

Richard McVay, 50, 632, 3000, 60, 159

Andrew Lilly, 20, 363, 700, 20, 358

Alexander Halsted, 70, 530, 1800, 75, 550

Andrew Roles, 60, 300, 300, 12, 628

Joshua Roles, 60, 740, 800, 4, 297

John Lilly, 30, 130, 300, 5, 270

Alexander Roles, 40, 360, 500, 5, 296

James Gore, 100, 500, 1000, 3, 787

Robert C Lilly, 60, 340, 3000, 15, 282

Denison Hale, 40, 110, 1000, 5, 211

John R. Moomaw, 60, 200, 2500, 150, 467

John G. Muncer, 120, 600, 5000, 200, 539

James Goodall, Renter, Renter, -, 5, 212

John S. Ewart, 200, 29800, 20000, 116, 882

Andre J. Hull, 100, 160, 2500, 10, 321

Henry Hull, 31, Renter, 300, 120, 280

William Chambers, 90, 135, 1790, 35, 286

Thomas Warden, 140, 1360, 3000, 40, 693

William O. Hollinsworth, 30, 170, 1500, 10, 235

Spurriel Baily, 100, 100, 2000, 40, 116

William Godby, 10, 142, 1000, 20, 191

Cyrus Snuffer, 100, 500, 300, 100, 1411

Randolph County, West Virginia
1860 Agricultural Census

The University of North Carolina at Chapel Hill filmed the 1860 agricultural census for Randolph County from originals at the West Virginia State Archives under a grant from the National Science Foundation in 1963.

Columns 1, 2, 3, 4, 5, and 13 represent the following information on the census:
1. Name of Owner, Agent or Manager of Farm
2. Acres of Improved Land
3. Acres of Unimproved Land
4. Cash Value of the Farm
5. Value of Farming Implements and Machinery
13. Value of Livestock

John Stalnaker, 90, 50, 2000, 50, 200
W. L. Daniels, 60, 130, 1000, 75, 400
Jonathan Daniels, 180, 300, 7500, 80, 300
Jacob Daniels, 30, 30, 520, 25, 300
Geo. W. Stalnaker, 100, 850, 4000, 20, 400
J. S. Stalnaker, 75, 50, 800, 10, 200
Emmet Buckley, 15, -, 705, 5, 200
Allison Daniels, 60, 99, 1500, 15, 70
Earle Daniels, 50, 113, 750, 25, 250
Madison Daniels, 400, 3000, 6500, 40, 2500
Jacob Daniels, 136, 170, 4000, 150, 550
Sol. W. Daniels, 150, 1000, 4000, 300,.750
A. Weese, 65, 50, 650, 25, 150
Ely Schoonover, 75, 225, 1500, 25, 250
Joseph Schoonover, 10, 233, 100, 60, 100
Marshall Schoonover, 30, 100, 600, 30, 40
Andrew Weese, 10, 20, 251, 20, 200
John W. Weese, 150, 50, 100, 50, 250
Ely Weese, 200, 200, 3000, 100, 700

Jacob Weese, 700, 50, 2500, 100, 400
Mary Weese, 20, 30, 300, 25, 150
John Marstiller, 50, 50, 1000, 50, 180
Sarah Chenoweth, 220, 100, 4000, 30, 750
Alba Chenoweth, 250, 210, 1500, 50, 400
James Chenoweth, 8, 200, 150, 4, 20
John Chenoweth, 100, 100, 2500, 12, 120
Job W. Daniels, 100, 1500, 1400, 15, 420
W. C. Webley, 100, 500, 600, 25, 300
Hiram Long, 25, 140, 400, 10, 45
E. C. Canfield, 35, 265, 800, 10, 192
Amos Canfield, 100, 700, 1500, 25, 335
John McQuane, 60, 240, 1200, 10, 192
Jacob Lanes, 150, -, 3000, 100, 760
Levi M. Wilmoth, 275, 2064, 6000, 50, 3325
John G. Currence, 100, 60, 1000, 100, 650
John Crouch, 120, 800, 10000, 150, 1500

Abraham Crouch, 120, 1000, 10000, 150, 1500

Jonas J. Simmons, 70, 930, 3000, 50, 350

Jacob Crouch, 1000, 3000, 30000, 150, 6000

J. W. Stalnaker, 45, 600, 2000, 50, 500

Aaron Coberly, 75, 400, 1500, 100, 200

Levi Coberly, 20, 180, 550, 15, 200

Jesse Coberly, 15, 81, 400, 5, 194

Acra Harisford, 30, 70, 100, 10, 150

Jacob Wilfong, 10, 100, 200, 5, 75

Joseph Summerfield, 15, 85, 200, 5, 15

George W. Wood, 100, 150, 4000, 100, 500

Ezra P. Hart, 30, 323, 200, 20, 100

F. Butcher, 300, 1200, 3000, 150, 400

L. M. Camden, 100, 100, 3000, 20, 400

R. J. Lambert, 18, 57, 600, 25, 100

A. Carper, 600, 360, 30000, 50, 4000

W. J. Long, 1000, 1100, 90000, 200, 6740

Mathew Wamsley, 125, 400, 2000, 125, 700

A. M. Wamsley, 200, 300, 15000, 250, 2000

Jacob Yokum Jr., 10, 150, 300, 20, 155

Jacob Yokum Sr., 100, 75, 2000, 150, 390

Peter Crickard, 60, 68, 800, 25, 380

Wm. Heter, 50, 78, 5000, 100, 450

Jonathan Wamsley, 80, -, 1000, 10, 135

Elias Yokum, 15, 100, 200, 10, 200

John Wamsley, 50, 50, 1000, 15, 150

John Smith, 100, 500, 3000, 100, 400

Catherine Crouch, 100, 80, 3000, 15, 550

Isaac Crouch, 300, 700, 10000, 25, 850

T. B. Scott, 750, 2700, 15000, 40, 1500

Wm. Scott, 30, 350, 1500, 20, 500

W. H. Currence, 150, 250, 10000, 100, 500

David C. Channel, 10, 80, 1000, 10, 115

Melvin Currence, 125, 175, 9000, 25,700

Tobias Long, 5, 395, 500, 15, 300

Abraham Heter, 700, 1000, 50000, 100, 2000

John A. Huter (Heter), 550, 3000, 25000, 200, 4000

Moses Huter, 500, 5000, 30000, 200, 4000

Jacob S. Wamsley, 750, 500, 15000, 150, 3500

Moses H. Crouch, 400, 300, 25000, 100, 4000

John M. Crouch Sr., 350, 800, 20000, 100, 3000

Isaac W. White, 30, 20, 1000, 20, 150

Jesse C. Ward, 600, 1300, 30000, 300, 5000

Jonathan Crouch, 200, 1450, 8000, 20, 375

Peter Couger, 20, 100, 500, 5, 20

Daniel Kalar, 4, 240, 500, 50, 289

David Salsbury, 30, 90, 500, 15, 200

George Hamer (Henner), 80, 105, 1800, 20, 250

Sol Hamer (Henner), 40, 60, 1500, 6, 150

Samuel Luecker, 25, 138, 100, 15, 100

Jacob Lundly, 30, 175, 2000, 25, 275

W. L. Riggleman, 20, 105, 400, 10, 160

Martin Riggleman, 70, 203, 2000, 10, 340

Silas Ramsey, 650, 4350, 8699, 12, 150

Peter Conrad, 15, 1085, 2000, 10, 150

Lewis Couger, 40, 960, 2000, 10, 170

W. H. Henner, 50, 30, 1000, 8,175

Wm. Folks, 50, 106, 1500, 15, 300

M. D. Ruckman, 100, 225, 2500, 20, 300

J. M. Wainum, 30, 4000, 5000, 7, 150

J. B. McCloud, 550, 200, 6000, 10, 40

A. Ward, 300, 200, 12000, 8, 150

Solm. Wamsley, 100, 169, 5000, 50, 500

Alexr. F. Stalnaker, 65, 80, 2500, 10, 125

Adam Timmons (Simmons), 160, 55, 3500, 150, 500

J. W. Moore, 100, 250, 5200, 75, 800

Thos. E. Wood, 23, 28, 220, 7, 175

Davis M. Wood, 10, 83, 280, -, 140

Joseph Moore, 500, 300, 6000, 50, 950

Augustus Wood, 60, 120, 4000, 75, 350

Samuel Wood, 100, 100, 2000, 50, 1000

Henson Douglas, 75, 165, 2000, 25, 550

John C. Wood, 15, 250, 1000, 10, 125

Paul Hamilton, 86, 140, 2200, -, 500

J. Q. Wilson, 30, 170, 2000, 75, 410

David O. Wilson, 50, 300, 1500, 35, 450

James D. Wilson, 50, 300, 1500, 40, 460

Amos Henner (Hamer), 750, 4250, 15000, 150, 2000

P. A. Tally, 35, 451, 200, 150, 300

Geo. W. Hare, 25, 200, 500, 8, 50

Alexr. C. Logan, 13, 187, 1000, 12, 225

John W. Logan, 25, 215, 1000, 15, 250

John D. Conrad, 140, 288, 8000, 15, 275

Wm. E. Logan, 30, 300, 2000, 10, 250

Hiram Ware, 100, 140, 1500, 8, 600

Elizabeth Channel, 40, 290, 400, 5, 60

M. C. Potts, 80, 25, 1500, 20, 800

Amos Wymer, 40, 160, 800, 10, 250

Sam. Channel Jr., 60, 115, 800, 10, 450

John Channel Sr., 100, 450, 5000, 20, 50

John N. Ware, 8, 21, 100, 5, 50

Jesse W. Simmons, 20, 80, 500, 15, 150

Moses Arbogast, 25, 45, 1200, 15, 250

Willis Taylor, 40, 60, 1200, 10, 225

Harman Snyder, 500, 1200, 15000, 100, 3500

Jacob Conrad, 200, 3000, 12000, 300, 750

Adam Lee, 66, 500, 5000, 20, 800

Alexr. Stalnaker, 60, 600, 3000, 12, 700

G. W. White, 30, 10, 500, 10, 180

Aaron Bell, 100, 100, 7000, 25, 800

J. Rosecrantz, 35, 318, 800, 20, 400

Wm. M. Wamsley, 200, 800, 12000, 10, 500

John M. Wamsley, 200, 800, 12000, 50, 450

M. W. White, 25, 65, 1000, 10, 80

Seybert White, 40, 60, 1000, 10, 100

Wm. Tacy, 14, 99, 500, 12, 100

J. W. White, 15, 35, 500, 10, 250

Benj. Kelly, 30, 120, 800, 10, 700

W. G. Ward, 700, 3000, 10000, 549, 4500

S. G. Mathews, 30, 170, 500, 10, 85

Jonathan Currence, 400, 1200, 7000, 20, 500

J. A. McCall, 70, 20, 5000, 10, 340

T. N. Snelser, 40, 600, 1500, 5, 190

N. C. Moss, 25, 112, 1000, 5, 75

W. D. Armstrong, 40, 700, 1500, 8, 200

Ely Crouch, 80, 400, 6000, 10, 350

Chas. Crouch, 100, 100, 4000, 20, 400

Jno. Shrere, 40, 660, 1500, 9, 180

Thos. Quick, 30, 130, 2000, 100, 360

Wm. P. Bradley, 70, 430, 3000, 100, 500

M. H. Bradley, 50, 280, 1000, 10, 210

A. B. Ward, 500, 1000, 10000, 175, 3700

Johnson Phares, 500, 1000, 30000, 28, 2105

Jesse F. Phares, 100, 600, 11000, 10, 125

Samuel Morrison, 100, 10000, 400, 50, 350

C. W. Butcher, 50, 2500, 2000, 75, 470

L. D. Currence, 30, 570, 1000, 20, 7

F. M. White, 150, 350, 4000, 150, 360

Isaac White, 100, 100, 3000, 20, 300

Mary Earle, 150, 235 10000, 75, 875

Michael Yokum, 50, 132, 1500, 20, 247

T. J. Capliner(Caplinger), 50, 950, 2000, 40, 400

Abraham Hinkle, 300, 150, 7000, 360, 850

Sol. C. Caplinger, 110, 445, 3000, 100, 600

George Caplinger, 75, 3000, 4000, 100, 374

George W. Caplinger, 50, 975, 3500, 100, 750

Jesse Harper, 140, 700, 3000, 100, 2119

Sarah Rowe, 100, 1400, 3000, 5, 25

Elijah Kittle, 120, 180, 3000, 40, 400

Thos. Scott, 120, -, 2000, 100, 400

Mary Curtis (Custis), 125, 210, 6000, 100, 417

Levi Ward, 275, 400, 4000, 200, 1500

E. R. Lough, 75, 75, 2000, 20, 940

Daniel Harper, 35, 65, 1600, 10, 400

Andrew Scott, 30, 168, 1000, 10, 200

Sol. Hinkle, 66, 10, 1000, 25, 180

Ananias Hinkle, 400, 150, 10000, 100, 1500

Wm. Foggy, 150, 205, 2000, 75, 400

Eliz. Stalnaker, 30, 140, 1400, 10, 200

George Scott, 50, 100, 1500, 20, 120

John Weese, 100, -, 1200, 20, 500

S. R. Scott, 50, 50, 1000, 20, 75

Geo. C. Litle, 100, 120, 4000, 100, 467

Wm. Isner, 100, 525, 2000, 100, 669

W. D. Currence, 100, 3126, 5000, 30, 795

J. M. Ball, 160, 100, 4000, 20, 570

Joseph Hart, 150, 2850, 8000, 50, 1300

J. S. Kittle, 10, 100, 100, 15, 100

P. T. Phillips, 20, 480, 250, 10, 75

Wash. Hilleany, 100, 900, 5000, 15, 200

J. W. Lewis, 60, 112, 2000, 25, 350

B. W. Kittle, 8, 84, 168, 25, 250

Patrick Raferty, 12, 88, 200, 15, 100

Jas. Nockton, 10, 75, 200, 10, 75

Job Staunton, 20, 230, 350, 5, 75

Geo. W. Mills, 80, 800, 1200, 30, 500

Wilson Osburn, 55, 69, 700, 20, 500

Alvin Osburn, 10, 90, 100, 10, 125

Jeremiah Lanham, 50, 50, 1000, 20, 220

Isaac Cutwright, 8, 42, 300, 10, 45

Alexr. Grim, 50, 100, 1200, 20, 150

A. G. Queen, 30, 75, 600, 10, 150

A. Markly, 4, 546, 1000, 20, 120

J. A. Wolf, 4, 26, 100, 5, 25

Preston Taylor, 15, 80, 600, 20, 50

Wm. Taylor, 28, 160, 1000, 20, 300

A. Nicholas, 16, 59, 300, 10, 100

Zach. Clellan, 4, 32, 100, 5, 40
E. Zicafoos, 15, 93, 500, 10, 200
Henry Zicafoos, 50, 700, 1000, 20, 300
H. W. Simmons, 15, 225, 700, 20, 200
H. W. Leigh, 20, 100, 350, 15, 150
John Light, 10, 200, 260, 10, 75
F. M. Light, 4, 61, 300, 5, 40
Solomon Wamsley, 4, 31, 100, 8, 75
Samuel Williams, 170, 30, 800, 20, 130
Dennis Demoss, 4, 47, 100, 5, 10
Addison Tinney, 6, 94, 350, 10, 35
Josiah Tinney, 12, 98, 150, 5, 120
John H. Westfall, 30, 90, 500, 20, 130
David Morgan, 50, 1050, 4000, 25, 620
Frederic Vanguilder, 10, 90, 500, 20, 300
Elijah Brain, 6, 295, 1200, 20, 300
Ere Mahhan, 140, 60, 4000, 25, 150
Catherine Warmer, 11, 223, 4000, 8, 120
Levi Smith, 40, 250, 2500, 20, 541
Jacob Herron, 26, 300, 1000, 15, 500
Samuel Channell, 100, 1400, 6000, 50, 700
J. C. Channell, 30, 670, 2000, 20, 200
W. M. Shiftlett, 12, 148, 150, 10, 25
J. A. Snelson, 31, 397, 400, 10, 175
Granville Lamb, 4, 700, 1000, 14, 70
Willis Herron Jr., 15, 285, 1000, 10, 120
Henry Harper, 330, 3670, 12000, 200, 1500
Wm. Clarke, 50, 50, 700, 25, 100
Levi D. Ward, 300, 600, 4000, 120, 1000
Gerah Everett, 20, 94, 400, 6, 65
Jacob Ward, 105, 700, 5000, 125, 1000
Scott Hill, 90, 800, 3000, 10, 350
Wm. L. Hill, 150, 200, 6000, 25, 800

Zines Weese, 250, 750, 8000, 135, 2200
Wm. F. Corley, 80, 220, 2000, 20, 450
Wm. Clem, 130, 70, 1800, 20, 250
Michael O'Conner, 40, 60, 500, 15, 120
James Williams, 75, 250, 2000, 20, 360
Elias Coffman, 30, 70, 500, 10, 200
Wm. H. Hiliard, 20, 77, 175, 8, 75
Wm. Hays, 45, 40, 400, 10, 160
George Hays, 20, -, 100, 6, 180
John Coberly, 30, 100, 800, 15, 200
C. Whitecotton, 25, 125, 1000, 10, 178
Elias Coberly, 15, 125, 300, 12, 50
Abel Pharis, 40, 360, 800, 20, 300
Archd. Wilson, 300, 562, 4000, 25, 450
J. K. Scott, 30, 270, 1000, 20, 400
Crawford Scott, 120, 180, 3000, 25, 508
John Nallin, 18, 82, 600, 10, 230
Pat. Flanegan, 25, 113, 200, 8, 110
Morris Henophen, 15, 135, 200, 6, 100
Pat. O'Conner, 25, 75, 250, 10, 150
Oliver Scott, 12, 100, 200, 8, 50
Levi Finley, 40, 125, 1600, 15, 350
Moses J. Phillips, 20, 462, 2600, 50, 200
Pat Durkin, 20, 80, 400, 5, 60
Geo. Phillips, 15, 130, 300, 8, 50
Jonas Lantz, 4, 46, 100, 3, 50
Pat. King, 10, 190, 200, 5, 80
Michael King, 20, 180, 200, 8, 150
Miles King, 10, 90, 100, 3, 25
Owen Riley, 14, 86, 200, 6, 80
Andrew Durkin, 20, 30, 200, 5, 75
James Brooks, 40, 70, 500, 6, 100
Luke White, 30, 70, 300, 5, 125
Elizabeth Apperson, 14, 14, 100, 5, 130
Ed. S. Tolbert, 50, 50, 500, 20, 325

W. C. Proudfoot, 35, 65, 400, 10, 130

Danl. Taughterry, 45, 205, 700, 8, 135

Jno. O'Donnell, 50, 350, 600, 40, 120

Cyrus Kittle, 30, 230, 2000, 15, 220

John Scott, 50, 400, 2500, 15, 200

Andrew Scott, 40, 260, 2000, 5, 225

Whitman Ward, 350, 2000, 10000, 60, 600

James Powers, 14, 144, 300, 30, 100

Chas. Suedia, 40, 75, 800, 20, 175

Levi S. Ward, 100, 265, 2000, 30, 367

J. M. Hart, 50, 84, 1700, 15, 350

Jacob M. Weese, 600, 900, 9000, 200, 2513

J. Cunningham, 15, 85, 400, 10, 30

Adam Mouse, 95, 63, 3000, 100, 350

Benj. I. Phares, 75, 325, 2000, 25, 260

Johnson I. Stalnaker, 50, 78, 900, 40, 160

Levi W. Stalnaker, 60, 110, 1500, 25, 300

Elias W. Pharis, 35, 765, 1000, 125, 250

John W. Ward, 30, 370, 1500, 20, 400

Zebn. Stalnaker, 25, 118, 700, 30, 125

Wm. Prirey, 50, 50, 800, 20, 150

Jas. H. Hicks, 20, 90, 600, 10, 250

Lorenzo Denton, 11, 89, 600, 9, 180

Wm. Folsbury (Folsburg), 6, 4, 150, 5, 125

Jonathan Workman, 50, 50, 600, 20, 250

Jas. Wilmoth, 75, 85, 2000, 25, 400

D. C. Wilmoth, 140, 200, 2000, 10, 400

John Wilmoth, 80, 50, 1500, 15, 350

Chas. W. Burke, 50, 106, 1000, 10, 325

Nicholas Wilmoth, 50, 20, 800, 150, 230

Isaac Wilmoth, 20, 5, 500, 150, 220

B. F. Wilmoth, 75, 150, 2500, 100, 300

Ellis B. Vanscoy, 30, 70, 80, 15, 56

Wm. Vanscoy, 25, 145, 400, 10, 180

Benj. J. Wilmoth, 75, 35, 500, 5, 90

D. M. Carrick, 30, 70, 200, 5, 40

Wm. Channell, 30, 69, 600, 20, 100

John Rice, 15, 85, 400, 20, 170

Hickman Chenoweth, 100, 372, 3000, 125, 800

J. C. Skidmore, 100, 133, 2000, 200, 600

Alpheus Skidmore, 200, 100, 3000, 200, 900

Daniel Canfield, 20, 170, 500, 20, 25

Samuel Canfield, 30, 150, 300, 10, 60

Delila Vanscoy, 50, 90, 1000, 15, 150

J. W. Wilmoth, 75, 425, 3000, 75, 400

Wyatt Ferguson, 70, 830, 2000, 25, 160

Dolbeau Kelly, 100, 120, 2500, 20, 320

Henry Harris, 75, 400, 2000, 15, 350

Jacob Vanscoy, 55, 100, 1000, 20, 250

D. S. Gainer, 50, 135, 1500, 50, 375

S. H. Gainer, 25, 75, 800, 10, 160

Samuel W. Gainer, 3, 97, 800, 5, 115

Melton Triplett, 5, 378, 300, 10, 50

James Murphy, 7, 43, 100, 5, 100

Geo. N. Martin, 10, 90, 400, 8, 25

R. G. Gainer, 100, 100, 3000, 30, 300

J. L. Schoonover, 100, 300, 3000, 25, 981

Thomas Schoonover, 150, 350, 3500, 25, 1450

Elizabeth Schoonover, 75, 125, 2000, 12, 230

Ely Smith, 45, 305, 2000, 15, 330

Edmund Wilmoth, 60, 340, 2000, 75, 350

Taylor I. (T.) Wilmoth, 50, 80, 800, 20, 120

Wm. C. Dizard, 60, 20, 800, 20, 320

Daniel Hart, 45, 150, 1000, 30, 220

Geo. W. Wilmoth, 40, 50, 800, 10, -

D. B. Hart, 37, 50, 500, 12, 120

Jacob S. Hart, 37, 50, 500, 15, 100

L. D. White, 125, 150, 3000, 25, 700

Geo. Allender, 10, 90, 100, 10, 180

Geo. W. Rennix, 12, 88, 600, 18, 25

D. A. Murphy, 45, 105, 1200, 20, 180

Elihu Wilmoth, 18, 35, 1000, 15, 120

Solm. Ferguson, 50, 200, 1500, 25, 410

Archd. Wilmoth, 30, 45, 1000, 20, 142

J. M. Pharis, 145, 855, 3000, 30, 850

James Vanscoy, 35, 65, 500, 15, 250

Eliza Stalnaker, 35, 130, 1000, 10, 280

Wm. Bright, 25, 106, 500, 10, 130

Abel Hyre, 100, 600, 2000, 100, 576

G. D. Hyre, 100, 400, 2000, 100, 476

Everett Chenoweth, 100, 150, 1500, 40, 192

Asbery Stalnaker, 100, 500, 2500, 25, 400

Joseph Simmons, 300, 240, 3500, 100, 700

Abraham Hyre, 150, 300, 3000, 40, 675

Samuel Wilmoth, 35, 130, 600, 5, 120

Wash. Taylor, 200, 900, 5000, 20, 1433

Hyre Stalnaker, 8, 86, 500, 10, 120

Granville Stalnaker, 15, 85, 400, 12, 130

Daniel Dinkle, 400, 700, 8000, 200, 2130

Garrettson Stalnaker, 100, 274, 2000, 25, 510

Jacob Phares, 65, 400, 1500, 200, 238

George W. Phares, 45, 125, 1200, 200, 385

Anna Phares, 150, 150, 3500, 225, 576

Joseph Harding, 10, 115, 500, 15, 282

Jesse Phares, 75, 325, 2000, 20, 600

Nancy Martena, 150, 70, 3000, 30, 400

Phillip Watts, 30, 120, 500, 10, 100

A. J. Gorden, 30, 120, 900, 25, 225

John P. Coberly, 71, -, 900, 30, 190

Wm. H. Coberly, 175, 1500, 3600, 300, 631

Henry Isner, 100, 300, 900, 10, 80

Allen Isner, 50, 300, 700, 50, 170

John A. Rowan, 60, 90, 1000, 75, 350

George W. Rowan, 60, 63, 1000, 75, 310

Jas. H. Lambert, 35, 205, 700, 15, 250

Thomas Isner, 100, 700, 3500, 70, 360

Jacob Piercy, 20, 340, 300, 10, 200

Orlando Woolwine, 180, 466, 6000, 70, 1150

Moses Harper, 112, 235, 3500, 50, 1235

J. I. Chenoweth, 100, 500, 1500, 50, 483

John Hornbeck, 100, 163, 1600, 15, 300

E. J. Nelson, 30, 250, 300, 15, 150

Cyrus Isner, 40, 160, 400, 20, 165

Abel W. Kelly, 115, 165, 1500, 150, 190

Abel H. Kelly, 40, 300, 2500, 15, 150

Ephraim Triplett, 15, 35, 300, 10, 120

John Triplett, 40, 60, 600, 20, 140

Archd. Coberly, 50, 135, 500, 15, 134

Wm. J. Isner, 15, 112, 200, 20, 125
Job Triplett, 60, 1860, 3000, 30, 418
_____ B. Weese, 15, 1685, 800, 15, 200
Isaac Taylor, 60, 195, 800, 20, 335
John Wyatt, 75, 4925, 1500, 15, 140
T. J. White, 200, 900, 3500, 25, 520
J. W. Mullenix, 150, 1000, 2700, 20, 850
Aaron Armontrout, 150, 1400, 4000, 155, 1500
Thomas Summerfield, 60, 30, 700, 25, 110
Christian Cooper, 10, 55, 250, 20, 190
Samuel Cooper, 50, 295, 1500, 25, 190
Mary Cooper, 40, 190, 400, 18, 270
Thompson Elza, 40, 156, 500, 15, 120
R. I. Johnson, 50, 1150, 2600, 15, 1920
Sylvanus Carr, 15, 185, 350, 35, 80
Joseph White, 40, 122, 900, 25, 300
Henry J. White, 35, 40, 700, 15, 260
Levi White, 40, 60, 800, 35, 278
John Pennington, 50, 392, 1200, 25, 476
John Snyder, 40, 1600, 2000, 40, 1200
Sol. A. Pennington, 45, 526, 1800, 20, 800
Labon V. Smith, 45, 320, 1200, 15, 360
V. B. Pennington, 35, 315, 400, 10, 270
Archd. Bonner, 50, 300, 1200, 22, 400
Jacob Roy, 60, 740, 2000, 25, 410
Wm. White, 50, 413, 2000, 12, 550
John Taylor, 140, 6200, 14000, 75, 1350
J. M. Crouch Jr., 150, 285, 2500, 20, 600
Isaiah Isner, 12, 259, 450, 15, 120
John Leary, 18, 100, 500, 15, 100

David Blackman, 890, 2455, 25000, 125, 475
Samuel Crane, 120, 280, 3000, 95, 200
Hoy McLean, 180, 462, 7000, 150, 1051
Jonathan Arnold, 800, 50000, 75000, 100, 4000
Geo. Buckley, 115, 50, 4000, 87, 335
Squire Bosunth, 223, 55, 2000, 25, 340
J. H. Logan, 50, 8950, 8000, 25, 125
J. S. Collett, 125, 141, 2000, 40, 80
Ely Kittle, 37, 48, 3500, 150, 320
Wm. Hyre, 200, 294, 1500, 70, 500
Thomas Collett, 200, 799, 3000, 200, 250
E. H. Chenoweth, 71, 58, 500, 30, 488
J. W. Crawford, 300, 4800, 15000, 300, 3975
Archd. Stalnaker, 70, 113, 1200, 100, 600
B. W. Crawford, 200, 1685, 15000, 20, 1915
Abs. Crawford, 200, 433, 16000, 25, 2000
Wm. & J. B. Ryan, 83, 85, 3000, 70, 381
Benj. Phares, 300, 1690, 20000, 100, 2300
Henry Currence, 100, 287, 7000, 50, 500
Robt. A. Crawford, 200, 140, 9000, 75, 800
Samuel Wamsley, 750, 500, 15000, 150, 3500
Wm. Hamilton, 1000, 1000, 10000, 50, 1238
Chas. Morgan, 20, 80, 300, 10, 180
Andrew McCally, 100, 900, 2000, 20, 350
Wm. M. Pickens, 100, 650, 4300, 25, 280
Geo. W. Collett, 4, 98, 100, 5, 100

Wm. D. Ferguson, 20, 106, 950, 10, 150

J. W. Marshall, 600, 1800, 15000, 60, 3800

Hamilton Stalnaker, 70, 1330, 2000, 200, 1500

Geo. W. Chenoweth, 100, 65, 1000, 75, 500

Ritchie County, West Virginia
1860 Agricultural Census

The University of North Carolina at Chapel Hill filmed the 1860 agricultural census for Ritchie County from originals at the West Virginia State Archives under a grant from the National Science Foundation in 1963.

Columns 1, 2, 3, 4, 5, and 13 represent the following information on the census:
1. Name of Owner, Agent or Manager of Farm
2. Acres of Improved Land
3. Acres of Unimproved Land
4. Cash Value of the Farm
5. Value of Farming Implements and Machinery
13. Value of Livestock

F. Jos. Maze, -, -, -, 10, 120
Benj. Hall, 100, 200, 2200, 100, 436
Michael Carhegan, -, -, -, 5, 43
Wm. Long, -, -, -, 5, -
Danl. Pribble, -, -, -, 150, 259
Geo. H. Lemmon, 85, 222, 4605, 150, 546
Jacob Peckenpaugh, -, -, -, 10, -
Hiram P. Ayres, -, -, -, 10, 82
Thos. J. Lanham, -, -, -, 1, -
John C. Collins, -, 100, 130, 10, 20
Wm. Sharpnack, 80, 333, 4130, 200, 445
Hiram Sharpnack, 35, 70, 1500, 20, 230
Ch__. N. Morgan, 25, 75, 800, 10, 214
Saml. Bell, 50, 41, 1000, 50, 283
John Sharpnack, 74, 380, 4540, 82, 391
Danl. Tenant, 59, 100, 2000, 60, 450
Lewis Flahearty, 5, 65, 500, -, 15
Mary Buchanan, -, -, -, 1, 25
Isaac Smith, 23, 117, 840, 110, 208
Isaac Whiteman, 70, 130, 2000, 20, 206
John Bell, 47, 100, 1764, 20, 233
Lewis Louge, -, -, -, 5, 28
Danl. Eddy, 10, 86, 500, 10, 127
Saml. Tenant, 18, 79, 300, 10, 161

John Black, -, -, -, 5, 100
Fredk. Winerich, 50, 65, 1000, 20, 342
John Cain, 60, 20, 800, 5, 190
John K. Bradley, -, -, -, 8, 277
Danl. Donley, 17, 20, 400, 5, 79
Philip Deem, 130, 545, 5000, 100, 375
Ben. Philips, 100, 240, 5000, 100, 736
Richd. Rutherford, 200, 2600, 8900, 175, 2306
Archd. Rutherford, -, 150, 400, -, 191
Martin Dugan, -, -, -, 2, 5
Geo. W. Ogdon, 40, 67, 1500, 14, 326
David J. Cain, -, -, -, 10, 257
Jacob McKinney, 100, 375, 5000, 100, 788
Ann Wigner, -, -, -, -, 20
Sanford B. Carroll, 150, 317, 5000, 100, 472
John G. Skelton, ¾ (town lot), 275, 50, 20
Michael Shinley, 16, 90, 800, 6, 30
Noah R. Boston, 50, 42, 800, 75, 100
Jacob Hardbarger, 3, 73, 308, 50, 198

John McGinniss, 100, 800, 4500, 100, 194

John Carroll, -, -, -, 20, 258

William Ice, 40, 60, 2500, 40, 132

Wm. B. Lowther, 60, 110, 2550, 50, 182

Geo. Price, 40, 110, 1000, 30, 186

Danl. R. Wigner, 1, 49, 400, 10, 29

Barkus Ayres, 50, 1450, 5000, 30, 160

Elizabeth Martin, -, -, -, 10, -

Manley Collins, 20, 51, 213, 80, 130

Wm. Hall, 10, 90, 500, 20, 217

Nathan Kernes, 100, 1100, 3600, 150, 528

Henry J. Jackson, 1000, 11222, 36666, 509, 2445

Jacob Hatfield, 190, 120, 7000, 340, 870

Jonathan H. Haddox, 60, 45, 1500, 30, 150

Danl. Cokeley, 70, 716, 3700, 50, 180

Isaac Cokeley, 65, 121, 2500, 40, 694

Geo. Moore, -, -, -, 5, 116

John S. Hall, 32, 300, 2656, -, -

Henry Moates, 150, 1000, 2000, 100, 487

Wm. J. Moates, -, -, -, 16, 118

Catharine Smith, -, -, -, 5, 27

John Layfield, 125, 435, 2070, 100, 342

James Layfield, -, -, -, -, 59

James Mahon, -, -, -, -, -

Cornelius G. Cain, 100, 65, 3000, 150, 963

Jos. Moates, 50, 175, 2000, 35, 290

J. F. Hadley's heirs, 100, 174, 3000, -, 63

Ephraim T. Hadley, -, -, -, -, 250

Stephen H. Hadley, -, -, -, 40, 285

Wesley S. Hennen, -, -, -, 2, 100

James Pew, 35, 83, 600, 30, 325

David McGregor, 25, 873, 2308, 50, 70

James Carpenter, 30, 20, 2500, 90, 197

Fredk. Mauck, 42, 92, 1000, 10, 143

Danl. Minney, -, -, -, -, 91

Abner H. Jobes (Jones), -, -, -, 10, 108

John Six, -, -, -, 10, 18

John Merritt, -, -, -, 5, 13

Hiram Oates, 30, 84, 1500, 30, 370

Humphry Mount, 10, 90, 500, 15, 133

Geo. H. Kester, 40, 152, 1500, 30, 325

James B. Humphry, -, -, -, 1, 65

Peter Dillon, -, -, -, 5, 28

Manuel Rowen, -, -, -, -, 24

Sinclair Smith, -, -, -, 100, 135

Wm. Howard, -, -, -, 5, 127

Peter Lee, -, -, -, 3, 25

Jessee Lee, -, -, -, 5, 3

F. W. G. Camp, -, -, -, 2, 20

David Wareham, ¾, -, 300, 3, 25

Hazlett & Co., 62, 62, 1800, 405, -

B & O Railroad & N& Western VA Railroad, 6, -, 4319, -, -

John Surtees, -, -, -, 3, -

Jos. Martin, 5, 123, 2000, 25, 203

John Surtees agt for Price, 30, 814, 8440, -, -

James Tate, 30, 70, 1000, 10, 202

Absolom George, 20, 80, 1000, 10, 201

Elijah Heysham, -, -, -, 20, 64

Addison McCoy, -, -, -, 55, 250

John Nutter, 43, 100, 1716, 50, 170

Sarah Deem, -, -, -, 4, 43

James Webb, 60, 80, 1460, 85, 229

Magdalene Nutter, 2, 98, 500, -, 36

Mathew Nutter, 20, 80, 1000, 10, 79

Thos. Bathgate, 26, 174, 800, 13, 54

Geo. Rutherford, 125, 575, 5000, 150, 1280

Francis Nicholas, -, -, -, 4, -

Ezeriah Exline, 30, 50, 500, 10, 119

Saml. Ruckman agt for Susan Excline's children, 30, 93, 366, 10, 162

Wm. Elliot, 40, 86, 628, 15, 175

Felix Grayson, -, -, -, 6, 28

Sanford Neuzem, -, -, -, 20, 115

Rachel Slocum, -, -, -, 5, 100

Solomon Reed, -, -, -, 5, 81

Saml. Hibbs, 45, 155, 1600, 35, 294

Andrew Shields, 40, 60, 1000, 35, 375

John J. Rexroad, -, -, -, -, 22

Allen Neuzum, -, -, -, 5, 128

Margaret Smith, -, -, -, 5, 113

Martin Nash, -, -, -, 5, 119

Geo. _. Carder, 40, 60, 1000, 35, 322

Robt. Ross, -, -, -, 20, 147

Wm. Evrett, 75, 125, 1000, 35, 615

Robt. G. Armstrong, 20, 180, 800, 35, 190

Luke Terry, 50, 450, 1620, 50, 360

James H. Terry, 150, 1050, 6000, 206, 1040

Dr. A. J. Sangster (Langster), 10, 1990, 2000, -, -

Stephen Weekly agt. Jos. H. Terry, -, -, -, 3, 40

John Bumgarner, -, -, -, 60, 228

Saml. Hamilton, 15, 185, 1000, 10, 16

Wm. Hamilton, -, 200, 600, -, 80

Wm. Douglas, 150, 761, 4295, 81, 1013

John Douglas, -, -, -, 50, 160

Wm. Douglas, -, -, -, 10, 87

Esrom Arnot, 3, 145, 600, 5, 100

Jno. M. Satterfield, 10, 90, 500, 15, 80

Mary A. Satterfield, 40, 140, 1800, 5, 100

Seth T. Satterfield, -, -, -, -, 100

Bazil Hudkins, 10, 90, 800, 10, 254

Adam Cunningham, -, -, -, 10, 113

Hillery Pratt, -, -, -, 5, 63

John W. Ingram, -, -, -, 150, 405

Levi Hammer, 35, 161, 2000, 60, 232

James H. Cross, 50, 85, 2000, 10, 268

Lewis Propst, -, -, -, 2, 92

Barton Hudkins, -, -, -, -, 75

John Crawley, 80, 77, 4000, 30, 375

James Kelly, 30, 70, 800, 10, 213

Michael Broaderick, 30, 70, 800, 10, 204

Isaac Postleweight, 12, 46, 348, 5, 65

Richd. Ankrom, 25, 25, 500, 10, 134

Caleb T. Hamilton, 40, 110, 1200, 5, 106

Ephraim M. Butcher, -, -, -, 5, 105

Jacob Staly, 30, 63, 600, 5, 97

James Alkine, 75, 125, 2500, 130, 426

William Alkine, -, -, -, -, 154

James Cochran, 20, 35, 1000, 50, 210

Ephraim Martin, 70, 38, 2000, 53, 327

Henry Rollins, 40, 460, 3000, 60, 232

John B. Yeager, 50, 150, 2500, 15, 286

Thos. Hess, -, -, -, 5, 121

Chas. Surber, -, -, -, 5, 150

Nathaniel Parke, 12, 212, 1500, 5, 125

John G. Wigner, 60, 119, 2078, 30, 274

Jacob W. Wigner, 40, 60, 2000, 25, 331

Isaac H. Cunningham, 30, 345, 2000, 25, 278

Valentine Merritt, 8, 6, 280, -, -

Wm. F. Lowther, 40, 122, 810, 80, 446

Geo. W. Wells, -, -, -, 5, 125

Abijah W. Cady, ½, -, 400, 10, 29

Moses Star, 100, -, 1200, -, -

Lemuel Furr, -, -, -, 55, 244

Wm. Godfry, -, -, -, 25, 305

Jos. Stuart, 20, 42, 800, 30, 69

Margaret Stout, 18, 12, 500, 3, 117
Jonathan Smith, 40, 155, 1500, 40, 260
Abraham Smith, 50, 0, 2000, 30, 291
Nathaniel Mitchel, 40, 60, 600, 10, 42
Jesse Rolston, 18, 91, 436, -, -
Patrick Bennett, 15, 35, 250, 30, 142
Matthias Cook, 10, 17, 270, -, -
Sylvester P. Webb, 25, 380, 2025, 12, 119
Peter Kelley, 60, 162, 1400, 70, 282
Jeremiah Rolston, 15, 92, 1070, 30, 138
Joab Hyssan, -, -, -, 5, 125
John Rexroad agt. For J. Jackson, 150, 627, 8000, 30, 653
Wm. McGregor, 200, 289, 7335, 183, 676
Elias Butcher, 100, 100, 3000, 150, 640
Calvin Butcher, 15, 49, 200, 12, 200
James Whaly, 48, 100, 1600, 100, 301
Susan B. McGregor, 65, 139, 1500, 118, 345
Thos. M. Reed, -, -, -, 25, 200
Geo. Corbin, 25, 75, 500, 10, 110
Jno. Frankenberger, 13, 37, 500, 10, 94
John Weekly, 100, 100, 4500, 65, 603
Justus Weekly, -, -, -, 15, 180
Henry D. Martin, 80, 139, 2000, 40, 336
Madison Lambert, -, -, -, 110, 287
Thos. Pratt, 70, 51, 1500, 50, 386
Isaac Shreve, 65, 35, 1000, 10, 158
John Edgell, 60, 36, 1000, 20, 312
Isaac Edgell, 50, 150, 1800, 15, 151
Jos. M. Wilson, 40, 1115, 500, 10, 165
James Hainor, 50, 81, 1310, 5, 246
Jos. M. McKinney, 15, 31, 4600, 15, 200
Geo. W. McKinney, 12, 88, 500, 3, -

Calvin M. McKinney, -, -, -, 8, 58
Jno. W. Sammington, 50, 150, 2000, 30, 200
John T. Lacy, 188, 189, 4100, 120, 435
James M. Lacey, -, -, -, 10, 116
Wm. J. Piles, 30, 33, 800, 20, 334
David Musser, 50, 37, 1000, 15, 260
Wells Ricks, 40, 63, 800, 20, 248
Wm. Campbell, 167, 300, 4670, 85459
John P. Stuart, 20, 218, 1666, 70, 260
James Hague, 65, 185, 2500, 190, 442
Saml. Lock, -, 65, 390, 5, 46
Thos. McCollock, 40, 75, 1000, 5, 87
James L. Cunningham, 20, 80, 1200, 115, 292
Dennis Hayhurst, 6, 94, 700, 25, 150
Isaac Lambert, 154, 615, 7700, 100, 576
Ellen Williamson, 40, 260, 6000, -, 78
A. S. Cone, 150, 2850, 5000, 50, 375
J. J. Clutter, 20, 88, 1500, 20, 129
Wm. F. Boehm, -, 215, 1575, -, 60
John H. Rexroad, 30, 176, 1200, 10, 97
Wm. Rolston, 40, 60, 1500, 50, 260
John Culp, 100, 200, 2500, 108, 234
Wm. Culp, 40, 35, 800, 25, 60
Jos. Gorrell, 12, 87, 1000, 10, 118
Geo. Jones, -, -, -, 6, 120
Absolom Hulderman, 20, 120, 1000, 5, 207
Elizabeth Lowther, 8, 67, 700, -, 20
Mary Hall, 40, 68, 1080, 40, 500
John Douglas, 50, 65, 1200, 15, 200
Andrew Douglas, 73, 100, 2000, 45, 638
Andrew Young, 75, 100, 2000, 35, 278
Jacob W. Philips, -, -, -, -, 190
James Stuart, 40, 91, 1800, 10, 197

Catharine Campbell, 40, 460, 2500, 65, 273

Edwin E. Smith, 12, 228, 1000, 8, 60

Colin Welch, 25, 27, 520, 5, 75

Cummin Lee, 25, 27, 500, 5, 44

Jos. Pew, -, -, -, 8, 100

Ca__ Lowther, 32, 29, 600, -, 40

Richard Wanlass, 110, 354, 2162, 65, 934

William Hays, 30, 220, 1000, 10, 216

Lowther & Lake agt for Jas. Maban, -, -, 2000, -, -

Hannak Martial, 90, 1400, 3880, 75, 287

Saml. Moats, 150, 57, 1656, 75, 502

John Taylor, 85, 79, 2000, 20, 190

Lewis Six, 25, 115, 1000, 10, 97

Christian Douglas, 40, 60, 1000, -, -

William Roberts, 16, 44, 700, 25, 120

Andrew Cokley, 85, 130, 2500, 33, 287

John T. Smith, 75, 87, 4500, 100, 230

Danniel Boughenar, -, 170, 400, -, 146

William Patton, 90, 260, 7500, 25, 580

Cyrus Hall, -, 3760, 11148, -, 100

James Morris, 40, 83, 1600, -, -

Wm. M. Patton, 100, 400, 3000, -, -

Smith C. Hall, 100, 5900, 12000, -, 100

John Harris, 150, 77, 5000, 50, 341

Moses S. Hall, 10, 269, 500, -, 269

John Stars (superintendent of poor farm), 80, 80, 2500, 60, 287

Robert C. Kersheval, 45, 115, 1600, 71, 244

Z. M. Peirpoint, 85, 2785, 3737, 231, 500

William Moats, 128, 72, 5000, 70, 322

Wm. H. Douglas, 16, 98, 1400, -, -

Wm. Meredith, 75, 125, 1700, 20, 216

Solomon Hawkins, -, -, -, 70, 125

Isaiah Wells, 65, 768, 9275, 45, 304

John Carnal, 15, 67, 700, 10, 134

Joseph Lambert, 70, 77, 2000, 70, 251

John Knight, 70, 70, 1800, 25, 151

Thos. Curry, 25, 83, 1200, 62, 225

Benj. B. Patton, 40, 60, 1000, 15, 150

Wm. Knight, 4, 96, 500, 90, 338

Mary Miller, 25, 25, 400, 15, 108

Enoch Legett, 25, 25, 4000, 40, 243

Ann Harris, 100, 70, 2500, -, 76

David C. Norris, -, -, -, 40, 157

Remembrance Blue, 40, 584, 2500, -, -

Henry J. Rexroad, 65, 72, 1800, 65, 268

James Malone, 100, 825, 7000, 75, 564

James H. Rider, 60, 57, 1170, 40, 204

Noah Rexroad, 140, 107, 3500, 50, 419

John P. Harris, 140, 145, 5700, 262, 828

James McKinney, 45, 770, 3750, -, 100

John Trainer, 17, 96, 1000, 13, 123

Jacob Wigner, 35, 40, 800, 10, 108

William Wigner, 30, 25, 800, 10, 168

John Price, 15, 110, 600, 10, 98

John McMullen, 10, 90, 600, 5, 55

Franklin Moore, 100, 290, 4000, 71, 263

William Moore, -, -, -, 5, 84

Isaac Minor, 25, 27, 725, 5, 49

Wm. Carpenter, 50, 100, 1500, 210, 477

Frederick Tanner, 50, 200, 2500, -, -

Fielding A. Pratt, 25, 103, 384, 5, 102

John Rowson, 200, 500, 5000, 10, 269

Wm. H. Cunningham, 12, 42, 400, 10, 92

Thos. Stephens, 50, 50, 1200, 20, 284

Wm. Mahaney, -, -, -, 5, 228

Jeremiah Mahaney, -, -, -, 10, 224

John Mahaney, 70, 30, 500, 35, 151

Saml. Rowson, 7, 193, 1000, 5, 92

John Corban, 43, 201, 2000, 74, 251

Henry Fowler, 20, 60, 700, 5, 234

Benj. Wrick, 3, 97, 500, 5, 242

Thos. K. Rawson, 10, 90, 500, -, -

Elijah Cunningham, 30, 10, 800, 55, 325

Jo. H. Hughs, 3, 75, 450, 5, 50

Clark C. Horner, 40, 157, 1200, 5, 233

William Elder, 15, 27, 210, 25, 271

Wm. D. Lambert, 61, 100, 2000, 90, 369

Edith D. Martin, 300, 753, 11050, 150, 108

Marshal M. Martin, -, -, -, -, 465

Isaiah Collins, 50, 133, 1316, 10, 126

Barton Hickman, 25, 75, 800, 35, 135

Larkin S. Sill, -, 85, 340, 30, 200

Edmund J. Jarvis, 145, 664, 4952, 97, 332

Jno. K. Cunningham, -, -, -, 20, 148

Lafayette Heflin, 16, 121, 800, 10, 58

Jno. A. Jones, 18, 121, 1000, -, -

Larkin Cunningham, 5, 21, 300, -, -

Wm. Rush, 50, 138, 1600, 10, 180

Josiah C. Garner, -, -, -, 20, 272

Simon Lantz, 115, 644, 9108, 160, 1101

Wm. Denning, 20, 110, 650, 20, 178

Jno. Collins, 500, 1614, 18110, 165, 3013

Jacob C. Lantz, 80, 200, 2064, 77, 985

Saml. Thomas, 80, 389, 5000, 103, 927

Mary A. Thomas, 25, 175, 1600, -, -

Michael Hollem, -, -, -, 110, 258

Edmund B. Jones, 3, 97, 500, 5, 92

Levi Hess, 80, 270, 4000, 160, 237

Eli Cline, -, -, -, 20, 355

Wm. C. Haymond, -, 100, 200, 3, 186

Notley Willis, 50, 150, 2000, 10, 402

Isaiah Doke, 8, 66, 300, 3, 100

Nimrod Stull, 50, 150, 1500, 5, 129

Enoch Stull, 5, 170, 1000, 18, 45

Henry B. Collins, 100, 600, 3000, 45, 410

David Ridgway, 12, 88, 500, 5, 112

Danl. Haymond, 100, 430, 6625, 55, 764

Thos. Griffin, 50, 73, 1500, 25, 294

David A. McGinnis, 56, 100, 2000, 22, 349

Jacob C. Jones, 22, 52, 800, 10, 173

Mansfield H. Jones, 10, 75, 500, 5, 35

Thos. McMullin, 60, 90, 2000, 21, 111

Danl. S. Vancoat, 20, 230, 2000, 5, 215

Elisha Cottrill, 35, 138, 2000, 80, 270

M. M. Hitchcox, 55, 862, 2000, 5, 22

Wm. Hitchcox, 80, 319, 5124, 100, 477

Wm. L. Hitchcox, 10, 305, 2000, 5, 4

Peter Reed, -, -, -, 5, 155

Jos. F. Cox, 20, 265, 2500, 20, 234

Mathew Riggs, 100, 69, 3000, 92, 544

Elias Marsh, 300, 679, 5241, 136, 1657

Jacob C. Collins, 10, 90, 1500, 5, 157

M. D. Bartlett, 34, 99, 1000, 5, 125

Elizabeth Jones, 15, 145, 225, 5, 120

Jno. Ferrebee, 3, 107, 660, 5, 25

David Cox, 30, 418, 2000, 21, 89

John Nash, 25, 75, 800, 5, 133
John Garner, 80, 386, 3000, 65, 339
Thos. Dodson, 14, 198, 1000, 10, 28
Benj. Dodson, 40, 141, 2000, 40, 225
Elias Gregg, 65, 60, 1800, 48, 136
Robt. J. Leedum, 6, 48, 500, 3, 111
Jos. Dodson, 40, 50, 1000, 7, 85
Ruth Dodson, 100, 250, 3500, 25, 288
James French, 2 ½, -, 100, 2, 15
James A. Heaton, -, -, -, 23, 133
James A. Heaton agt. For J. Taylor, 75, 237, 3000, -, -
Benj. F. Heflin, -, -, -, 5, 42
Benj. F. Heflin agt. For D. M. Harris, 50, 100, 1000, -, -
Jos. Cunningham, 125, 225, 5250, 90, 740
Wm. Collins, 90, 972, 3652, 105, 600
Silas Sigler, 60, 34, 1500, 22, 141
Isaac Leedum, 20, 93, 2000, 60, 175
Barton Hickman, 20, 81, 1000, 41, 140
Eli M. Marsh, -, 3, 15, 35, 185
Enoch Marsh, 100, 125, 4000, 85, 1081
John H. Warner, 35, 10, 1000, 55, 158
Zelphia Rynehart, 175, 105, 3000, 75, 533
John Conwell, -, 127, 191, -, 41
Thos. T. Flinn, 50, 50, 600, 43, 390
John Flinn agt. for E. B. G. Reed, 75, 525 4800, -, -
Henry Howard, 35, 100, 1000, 40, 170
Mary Taylor, 60, 173, 2000, 40, 261
Edmund Taylor, 400, 800, 12400, 5, 179
Wm. H. McKinley, 40, 60, 1700, 3, 165
Rawley Haddox, 10, 19, 500, -, -
Philip Sigler, 7, 190, 1000, -, 25
David Dixon, -, 8, 16, 10, 17
Henry S. Boyce, -, -, -, 3, 40

Emily & Martha Peterson, 3, 23, 450, -, 25
Chas. Meserve, -, -, -, 60, 268
A. C. Barnard, 40, 166, 2000, 50, 184
Archd. J. Wilson, 380, 3015, 20792, 70, 3052
Andrew J. Harris, 20, 65, 800, 40, 102
Adonejah Watson, 75, 129, 2424, 80, 342
Isaac S. Cox, 65, 85, 1400, 15, 64
John Woodsides, 3, 3, 60, 5, 279
Johnathan Woodsides, 30, 120, 1200, -, -
Anderson Patton, 40, 146, 2000, 55, 426
Barton H. Wilson, -, -, -, 70, 317
Jerome B. Myres, -, -, -, 65, 245
Josiah A. Woods, 20, 80, 800, 15, 98
Tarleton Peck, 20, 75, 500, 6, 53
Geo. Stuart, 12, 13, 200, 5, 88
Ezekiel Shepherd, 20, 80, 800, 15, 111
Samuel R. Jones, -, -, -, 10, 70
Elias Richards, 10, 86, 500, 8, 118
Eli Riddle, 12, 168, 1000, 29, 30
David J. Riddle, 60, 132, 1000, 10, 63
John C. Riddle, 9, 102, 1000, 10, 40
Saml. Calhoun, 20, 89, 800, 11, 106
Wm. Dixan (Dixon), 15, 85, 1000, 10, 217
John M. Wilson, 100, 530, 5286, 58, 484
Solomon L. Dodson, 90, 400, 6000, 115, 566
Anson S. Merrifield, -, -, -, 10, 150
Edmond R. Taylor, 10, 344, 3540, 30, 448
Richard M. Taylor, -, -, -, 15, 444
Chas. R. Brown, 40, 140, 1500, -, 102
Levi S Wells, 81, 105, 400, 3, 38
John M. Wilson, 50, 132, 1500, 23, 310

John Shinn, 10, 45, 500, 5, 65
Syelus Hall, 60, 215, 2500, 66, 340
Benj. Richards, 40, 56, 1200, 10, 132
Geo. W. Richards, 25, 55, 500, 3, 164
Benj. F. Richards, -, 150, 300, -, 70
Jeremiah Fluharty, 40, 75, 920, 20, 136
Isaac Elliford, 15, 60, 500, 3, 32
Arjelon P. Myres, -, -, -, -, -
James Marquis, 30, 70, 600, 10, 94
James Davis, 5, 25, 200, 3, 43
Wm. Cook, 30, 636, 2500, -, -
Jo. Flanagan, 87, 200, 4000, 150, 205
John W. Dodson, -, -, -, 25, 219
Hiram Dodson, -, 30, 75, 12, 532
Manuel Dodson, 150, 280, 5500, 3, 334
Saml. Wiseman, 121, 88, 500, 10, 30
James Heatherly, 2, 188, 760, 10, 29
Thos. Pool, 40, 160, 3000, 55, 204
Esquire Dodson, 30, 130, 1000, 20, 63
Jefferson Broadwater, 200, 400, 1400, 115, 963
John W. Davis, 6, 68, 600, 5, 105
John Cross, 100, 173, 5500, 80, 648
Wm. W. Wilson, -, -, -, 15, 106
Thos. Hamrick, -, -, -, 7, 91
Susan B. McDouglas, 100, 120, 3500, 45, 964
Jacob Richards, 75, 230, 2610, 47, 493
Wm. W. Elder, 15, 135, 750, 3, 117
John Elder, -, 100, 500, 10, 292
Wm. H. Mincree, 20, 80, 1000, 5, 140
Rodger D. Davis, 30, 318, 1000, 28, 182
Jeremiah Snodgrass, 25, 129, 1000, 55, 100
Fanny Broadwater, 100, 200, 4500, 5, 174
James Woods, 100, 245, 2738, 60, 503

Benj. Wells, 15, 10, 3000, 55, 384
Margaret Cunningham, 25, 25, 800, 3, 72
Patsy Wells, 70, 30, 1500, 40, 134
Danl. Wigner, 1, 49, 400, 15, 28
Chas. D. Williamson, 30, 20, 1000, 21, 189
Philip A. Woods, 35, 21, 800, 5, 167
Elijah Wigner, 10, 40, 400, 3, 67
Wilson Patton, 17, -, 4000, 65, 119
Wm. Mealey, 48, 83, 786, 20, 169
Thos. Mealey, 40, 60, 2000, 15, 89
John Frieadly(Friendly), 30, 220, 1500, 25, 173
Nimrod Cross, 100, 100, 3000, 169, 248
Geo. W. Cross, -, -, -, 10, 81
John Cross, -, -, -, -, 112
Jacob Hayhurst, 53, 147, 2000, 43, 353
Thos. Mason, 65, 41, 1200, 33, 311
Wm. K. Elder, 40, 56, 1000, 55, 155
Martin Cochran, 60, 80, 2000, 90, 346
John Leggett, 60, 90, 2000, 45, 270
Wm. I. Lowther, 150, 266, 7320, 125, 833
Saml. J. Prunty, 44, 200, 2440, 30, 271
Jonathan McKinly, 200, 302, 8000, 110, 323
Thos. M. McKinley, -, -, -, 70, 416
Bingham Wood, 50, 45, 1000, 24, 385
Reason H. Wilson, 35, 65, 1000, 5, 38
Nathan J. Snodgrass, 75, 110, 1505, 22, 256
Stephen Clayton, 30, 44, 750, 30, 105
Elijah Clayton, 80, 370, 2500, 55, 405
Elias Snodgrass, -, -, -, 5, 77
Isaac Snodgrass, 30, 20, 450, 3, 23
Geo. M. Amos, 50, 54, 1000, 10, 148

Wm. B. Baker, 100, 153, 2024, 32, 617

Thos. C. Baker, 40, 60, 1000, 120, 190

Wm. B. Lowther, 75, 348, 3000, 110, 623

Robt. J. Wilson, -, 100, 200, -, -

Edward Lowther, 25, 75, 600, -, -

Bushrod W. Lawson, 30, 79, 1200, 40, 192

Eliza Lawson, 25, 75, 1000, 20, 175

John Hawkins, 65, 192, 2500, 33, 229

Jackson Shuttlesworth, 30, 70, 1000, -, -

Edward Rollins, -, 100, 300, 5, 43

John Cook, 130, 581, 5666, 80, 369

Olley Mancer (Maneer), 85, 121, 2530, 12, 496

Henry Hawkins, 50, 195, 2440, 96, 225

John Upton, -, -, -, 8, 130

Saml. Sinnett, 25, 25, 600, -, 182

Alexr. Lowther Jr., 150, 1385, 6769, 140, 1291

Felix Prunty, 100, 233, 3000, 108, 957

Alexr. Lowther, 100, 300, 6000, 55, 161

Joseph Hess, -, -, -, -, 130

Robt. W. Lowther, -, -, -, -, 201

Wm. S. Wilson, -, -, -, -, 20

John A. Lowther, 200, 1704, 7866, 155, 129

Mortimore Ireland, 55, 200, 3060, 5, 220

Reason Cain, 27, 100, 1700, 23, 78

Isaac Trimble, 60, 270, 2970, 32, 204

John Parker, 25, 35, 600, 5, 56

Marshall A. Neal, 20, 149, 1400, 50, 137

Eber Mason, 25, 100, 1000, 3, 81

Danl. Nay, 70, 48, 2000, 40, 289

James R. Jones, 60, 87, 1470, 47, 365

Isaac Valentine, 25, 25, 500, 15, 150

Sarah A. Maxwell, 16, 34, 500, 20, 115

Jo. Waggoner, 60, 20, 800, 20, 137

Isaac Gum, -, -, -, 5, 12

Wm. Stanly, 25, 31, 800, 10, 40

Jessee Cain, 100, 151, 2516, 67, 579

Geo. W. Troyman, 20, 132, 456, 10, 130

Jo. Kernes, 10, 95, 300, 3, 40

Horatio J. Mason, 50, 160, 1000, 25, 305

James Stanly, 7, 161, 336, 5, 29

Jo. W. Stanly, 12, 136, 290, 10, 67

Asa Kernes, 12, 123, 405, 5, 95

Chas. A. Kernes, -, -, -, 10, 164

Isaac Trader, 8, 103, 222, 3, 15

James Jenkins, 7, 20, 200, 3, 20

Madison Jenkins, 10, 50, 180, 5, 43

Andrew Jones, 5, 206, 422, 5, 20

Jacob Null, 12, 38, 250, 5, 65

Thos. P. Wilson, 10, 40, 250, 5, 53

John W. Mellan, 35, 236, 1000, 30, 269

Geo. Middleton, 25, 75, 500, 5, 126

Henry J. Pribble, 2, -, 1010, 25, 549

Hugh Pribble, 100, 202, 11740, 265, 770

Benona P. Leizure, 22, 64, 500, 5, 158

Isaiah Duff, -, -, -, 5, 103

James Deem 2nd, 20, 83, 350, 6, 123

Patrick Deem, 17, 46, 154, 5, 125

Benj. B. Nutter, 50, 87, 1200, 57, 235

Isaac Jackson, -, -, -, 80, 377

Wm. Williams, 40, 215, 1100, 15, 205

Michael Golden, 16, 109, 650, 10, 146

Henry L. Jackson, -, -, -, 75, 569

John L. Nutter, 40, 40, 1100, 15, 63

Peter Coil, 50, 100, 600, 23, 239

Patrick Coleman, 20, 105, 400, 5, 152

Barney Lyons, 20, 70, 600, -, -

Michael Lyons, 10, 90, 275, 1, 40
John O'Connel, 20, 80, 300, 3, 110
Fredk. Lemmon, 70, 2266, 6549, 95, 462
Alva H. Ayres, 25, 82, 642, 15, 193
John Ruckman, 50, 570, 4960, 10, 154
Asa S. G. Dilworth, 80, 156, 2360, 34, 240
Perviana Murphey, 36, 144, 2000, 65, 70
Eli M. Ayres, 35, 15, 1000, 12, 138
Edward Lough, 20, 80, 500, 5, 88
Simon Hickman, -, 100, 160, 5, 20
Wm. J. Drake, 20, 180, 1000, 20, 121
Elijah Williams, 30, 19, 500, 25, 258
John Stanly, 40, 183, 669, 10, 96
Thom. W. Stanly heirs, 28, 2, 200, -, -
James Dulaney, 48, 13, 800, 85, 105
John D. Senate, 35, 358, 2850, -, -
Abram. Cunningham, 30, 68, 800, 10, 113
Reuben Eye, 80, 280, 800, 58, 394
Zebulon Rexroad, 60, 140, 1200, 85, 302
Lewis Rexroad, 30, 213, 1000, 10, 179
Nicholas Swadley, -, -, -, 80, 240
Washington Mullennax, 1, 63, 192, 3, 30
Wm. Randell, 40, 94, 800, 50, 148
John Drake's heirs, 50, 136, 700, 20, 100
Solomon Mullennax, 30, 70, 600, 60, 147
Henry Layfield, 20, 110, 600, 8, 112
Absolem Cunningham, 7, 153, 400, 30, 130
Jacob Layfield, 25, 275, 600, 15, 250
Otho Zickefoose, 80, 313, 2924, 70, 447
Mary Smith, 80, 285, 2200, 5, 48
Benj. Waggoner, 10, 15, 300, 5, 112

Henry R. Senate, 50, 145, 2000, 20, 81
Jacob Cunningham, 6, 119, 1200, 20, 149
Patrick Drake, -, 322, 800, 80, 193
Elizabeth Drake, 50, 250, 1000, -, -
Jacob Hammer, 100, 380, 4000, 19, 778
Ephraim Cunningham, -, -, -, 5, 40
Thos. Williamson, 25, 68, 600, 30, 800
Peter Simmons, 100, 228, 3500, 115, 731
James Moyers, 42, 340, 1100, 53, 490
James Crummett, 40, 162, 1212, 90, 280
Thos. Hill Jr., 65, 138, 2003, 45, 171
Lewis Hammer, -, -, -, 5, 127
Wm. Webb, 62, 200, 4000, 167, 588
Henry Fulwider, -, -, -, 3, 23
Amos Jones, 14, 40, 500, 15, 249
Solomon Steel, 30, 88, 1000, 5, 110
Adam Cunningham's heirs, 45, 18, 1000, 5, 136
Peter Tharp, 40, 160, 1500, 10, 96
Saml. Zickefoose, -, -, -, 10, 169
Peter Zickefoose, 35, 49, 820, 10, 130
Thoms. Martin, 30, 130, 1040, 10, 120
James Webb, 30, 20, 400, 70, 125
James Star, 90, 4343, 3000, 88, 481
Wm. H. Jones, -, -, -, 10, 53
Wm. Jones, 40, 260, 2100, 100, 295
Wm. Jones agt. for Wm. L. Jackson, -, -, 1000, -, -
John Harris, 30, 70, 800, 25, 130
Wm. Cokeley, 50, 222, 2500, 41, 374
Solomon Rexroad, 30, 329, 1077, 33, 71
Jos. J. Kuykendall, 40, 107, 1000, 10, 212
Pamson Kendall, 10, 207, 3000, 45, 677

Jesse Morris, 35, 89, 620, 10, 253
Saml. Hammer, 20, 232, 1000, -, -
Henry H. Amos, 55, 125, 1800, 20, 287
James Brader, 15, 135, 500, 10, 35
Wm. Tibbs, 55, 75, 1000, 5, 208
Thoms. Harbargar, 45, 40, 570, -, 168
Harman Senate, 65, 157, 2500, 65, 417
Thos. Hagans agt., for Philip Wimer, 30, 70, 200, 2, 94
Aron Bell, 30, 192, 1300, 35, 182
Aron Kernes, 15, 25, 320, -, -
Nimrod Bennington, 20, 80, 800, 30, 300
Wm. P. Moates, 7, 93, 300, 10, 96
Abel Senate, 50, 150, 2000, 45, 147
Abram. Simmons, 30, 50, 300, 9, 254
Geo. Harold, 30, 159, 800, 12, 155
David Keek (Keck), 60, 130, 2000, 20, 157
Jesse Waggoner, 11, 14, 250, 5, 105
Wm. J. Cheatham, 40, 38, 800, 65, 240
John Reed, 85, 120, 2000, 10, 73
James J. Deem, -, -, -, 52, 335
Mary Clouse (C. Couse), 25, 13, 600, 10, 159
James H. Harris, 30, 32, 700, 100, 67
Peter Moyers, 144, 200, 4600, 120, 754
Geo. Sinnett, 120, 440, 5530, 303, 550
Jane Watson, 100, 100, 4000, 75, 558
Larkin Pierpoint, 40, 162, 1800, 50, 297
Daniel V. Cox, 180, 524, 6122, 120, 1033
Wm. L. Cox, 30, 48, 1000, 20, 355
Thos. Conway, 10, 90, 400, 15, 100
Elias Summers, 60, 90, 1200, 40, 357
James J. Prather, -, -, -, 5, 97

Ellis Hall, 25, 85, 990, 5, 125
Geo. Foster, 15, 100, 700, 5, 167
Icebud Kirkpatrick, 10, 90, 200, 3, -
Washington Parker, 40, 150, 1330, 5, 84
Hiram P. Cain, 30, 113, 852, 15, 138
Jo. S. Wilson, 50, 57, 2000, 68, 352
Wm. T. Mitchel, 30, 307, 842, 10, 297
Thos. D. Pritchard, 30, 59, 1335, 30, 388
Elijah Smith, 80, 193, 2000, 60, 613
Aaron Smith, -, -, -, 5, 179
James B. Westfall, -, -, -, -, 232
John Mitchel, 60, 372, 2200, 39, 436
Alpheus McDaniel, -, 58, 78, 30, 113
Riley Mason, -, -, -, 5, 118
Jonathan Baker, 25, 53, 400, 5, 156
Jo. Haddox, 45, 55, 1200, 41, 224
Harrison Wass, 10, 490, 1200, 22, 166
Eugenius Barker, 5, 90, 300, 7, 53
Valentine Butcher, 75, 72, 2205, 45, 500
Valentine Butcher Jr., 15, 113, 256, 5, 104
Geo. W. Butcher, 90, 95, 3000, 75, 270
Geo. Thomas, 55, 300, 3550, 10, 344
Alfred Malone, 15, 88, 824, 6, 35
Elias Thomas, -, 65, 400, 5, 121
Marshal Dean, 30, 139, 1500, 15, 185
Asa FitzRandolph, 50, 50, 1800, 40, 204
David Mitchel, -, -, -, 50, 51
Jeremiah Snodgrass, 90, 130, 3300, 65, 357
Elijah Snodgrass, -, -, -, 3, 112
Asa Bee, 2, 130, 1315, 2, 111
Henry H. Goodin, 2, -, 500, 5, 40
Joab Meredith, 60, 20, 2000, 35, 174
Ezebian Bee, 45, 29, 1700, 50, 500
Zebulon Bee, 13, 447, 2179, 5, 176
Lemuel Davis, 10, 121, 600, 8, 81
John Fox, 35, 100, 1080, 53, 126

Reba Davis, 20, 80, 900, 20, 176

Ezekiel Kelley, 45, 248, 1600, 54, 323

Josiah Bee, 14, 244, 1500, 25, 24

Jonathan C. Lowther, 25, 150, 1200, 24, 159

Reed E. Bond, 11, 59, 350, 3, 17

Harey J. Gumm, 14, 83, 500, 5, 50

Jeremiah Bee, 8, 20, 150, 3, 26

Jacob N. Ironsman, 80, 50, 270, -, -

Johnson H. Davis, -, 77, 400, -, -

Wm. & Zimy Flanaghan, 250, 550, 8000, 205, 1533

Henry S. Morris, 16, 197, 800, 5, 54

Peter T. Wilson and others, 75, 218, 2850, 60, 207

Mary Snodgrass, 75, 243, 2544, 83, 546

Wm. F. Ehret, 40, 64, 1000, 32, 252

Jacob Ehret, -, -, -, 2, 16

Benj. Prather, 40, 318, 1790, 20, 210

John L. Harris, 60, 58, 1180, 64, 379

Geo. Pritchard, 75, 300, 2500, 65, 451

Saml. G. Pritchard, 50, 60, 1760, 47, 156

Chs. W. Saterfield, -, -, -, -, 26

Peter Pritchard, 100, 300, 5000, 80, 557

Thos. W. Pritchard, 30, 91, 2000, 15, 135

Jackson J. Thompson, 30, 70, 750, 5, 101

Jacob Allender, 75, 250, 1800, 161, 366

Jacob P. Gecman, -, -, -, 5, 214

Geo. W. Zinn, 98, 100, 2200, 30, 538

Thos. Ireland, 100, 150, 1600, 80, 608

Robt. Ireland, 30, 100, 1300, 57, 314

Eliza Zinn, 40, 93, 1330, 10, 132

Quilly M. Zinn, 45, 55, 1500, 135, 371

Granville M. Zinn agt., for J. Girvin, 25, 1255, 2560, 13, 160

Christopher V. Nutter, 90, 910, 5000, 90, 542

Archd. Lowther, 50, 100, 3000, 155, 619

Hezekiah B. Tharp, 60, 240, 3500, 130, 340

Geo. Watson agt. of L. Maxwell, 20, 180, 400, 16, 261

John Watson for Sarah Watson, 4, 72, 304, 5, 117

Wilson Watson, -, -, -, 35, 218

John Jett, 40, 53, 930, 5, 406

Wm. Fox agt. of Lewis Maxwell, 200, 800, 8000, -, -

Selba Wade, 25, 79, 1040, 27, 66

Amos Pritchard, 1, 99, 650, 6, 149

Geo. W. Garrison, -, 100, 350, 5, 109

James D. Early, 5, 45, 225, 2, 30

James R. Brake, 40, 393, 2150, 80, 366

John W. Miller, 60, 140, 1600, 40, 182

Elisha Hall, 24, 222, 1500, 5, 188

Robt. Summerville, 200, 1450, 14850, 113, 713

Martin S. Summerville, 15, 85, 600, 20, 354

James W. Summerville, 40, 223, 669, 10, 452

John Goin, 20, 67, 300, 5, 117

Timothy Tharp, 160, 2903, 18834, 130, 1338

Ebenezer H. Tharp, -, -, -, 10, 361

Andrew Woofter, 25, 185, 1055, 5, 211

Wm. C. Hart, 3, 35, 150, 5, 57

Lemuel Hall, 125, 355, 3930, 112, 649

Lawson Hall, 10, 65, 400, 5, 100

Hiram Adams, 30, 70, 1000, 5, 115

David E. Brown, -, -, -, 20, 106

Isaac Haden, 90, 473, 5630, (120), 543

Andrew Law, 400, 529, 10000, 155, 2459

Evan Morehead, 20, 153, 800, 8, 130

Martin C. Ward, 120, 170, 2000, 105, 322

John Huff, 20, 380, 200, 15, 324

Asa Bee, 50, 350, 2000, 15, 425

Abigail Bee, -, 127, 381, -, -

Nancy Zinn, 50, 250, 2000, 10, 80

David F. Randolf, 1, 9, 20, -, -

Hezekiah D. Sutton, 20, 108, 650, 5, 110

Lucinda Lowther, 30, 30, 700, 10, 212

Wilson Prunty, 175, 360, 10000, 125, 1961

John P. Dulin, 130, 295, 4250, 72, 268

Wm. Harris, 100, 62, 3500, 135, 400

Robt. Lough, 200, 389, 9090, 98, 433

Robt. Means, -, -, -, 15, 186

John A. Prince, -, -, -, -, 88

Benj. Wolf, 218, 1311, 20590, 80, 510

Alexr. Glover, 50, 25, 1800, 80, 560

Morgan Rexroad, 80, 135, 1800, 65, 438

John M. Evans, -, -, -, 40, 314

Jacob Cunningham, 60, 233, 2500, 40, 90

Danl. Ayres, 40, 60, 1000, 10, 51

Zachariah Barker, 70, 42, 2000, 50, 186

Danl. H. Valentine, 50, 211, 1300, 73, 174

Jacob Wolf, 40, 260, 1500, 73, 200

Geo. _. Valentine, 78, 84, 2000, 53, 47

Jacob Daugherty, 12, 447, 800, 25, 96

Adam Harris, -, -, -, 10, 68

Henry Tingler, 100, 476, 4000, 40, 429

Julia Westfall, 7, 93, 300, 1, 67

Thos. Goff, 110, 133, 2500, 65, 392

Justice S. Goff, 12, 84, 400, 5, 83

Robt. Tibbs, 130, 270, 6000, 60, 393

Eugenius Tibbs, -, -, -, 20, 286

Thos. D. Tibbs, 50, 100, 1000, 75, 225

John Daugherty, 68, 100, 1500, 58, 422

John Schoolcraft, 30, 30, 800, 100, 308

Alexr. Goff, 104, 212, 2500, 40, 414

Henry Messenger, 40, 210, 9500, 25, 332

Jane Davidson, 12, 105, 1000, 8, 125

John Bird, 35, 140, 800, 20, 140

Benj. Goff, 200, 444, 3235, 25, 248

Jo. H. Goff, 40, 60, 600, 10, 105

J___ A. C. Davis, 40, 60, 500, 6, 38

Geo. Miller, 6, 204, 420, 5, 142

David McDonald, 20, 178, 500, 2, 40

Elizh. Strait, 8, 92, 600, 5, 45

Greenbury Hammon, 14, 351, 800, 5, 181

James Goff, 6, 94, 300, 5, 120

Salathiel G. Goff, 18, 85, 400, 5, 28

Thos. Newland, 12, 88, 400, 10, 84

Allen Calhoun, 40, 160, 1600, 48, 160

Riley Mason, 3, 97, 300, 5, 68

Alpheus Claton, 15, 185, 500, 5, 130

James Wright, 50, 148, 1800, 110, 333

James L. Smith, 16, 200, 1080, 25, 129

Harrison Wright, 100, 500, 4800, 32, 714

Harrison Bartlett, 50, 160, 2100, 10, 279

Josiah McDonald, -, 17, 85, 3, 18

Mc. McDonald, 36, 100, 680, 5, 143

Elijah Bartlett, 60, 135, 2100, 75, 377

Elmore Prunty, 10, 90, 500, 10, 106

Levi Smith, 80, 120, 2000, 97, 454

Jonathan Reed, 30, 70, 500, 20, 321

Ba__ J. Smith, 12, 112, 600, 5, 84

Lemuel Smith, 30, 100, 970, 5, 222

Nathaniel Smith, 24, 146, 1190, 5, 164

Davidson C. Riddle, 12, 88, 300, 5, 63

James F. Riddle, 5, 45, 150, 3, 89

Asby P. Law, 60, 450, 2000, 30, 259

Israel Davidson, 45, 55, 800, 8, 200

Levin Riddle, -, -, -, 10, 372

Horation N. Wilson, 6, 194, 1000, 5, 220

John C. Ireland, 25, 191, 864, 5, 50

Harvey C. Cutter (Clutter), -, -, 700, 5, 20

John W. Orendorf, 20, 180, 800, 5, 181

Wilson B. Cunningham, 10, 90, 400, 5, 29

Saml. Flemming, 15, 115, 780, 5, 51

Danl. Osbourn, 14, 41, 150, 5, 80

John Bollinger, 6, 94, 200, 3, 80

Lafayette Goff, 35, 152, 1000, 42, 472

John W. Goff, 30, 50, 500, 42, 177

Alexr. B. Goff, 25, 75, 600, 100, 148

James Beaden, 20, 80, 500, 10, 139

John Wass, 165, 539, 7000, 175, 724

William Collins, 50, 200, 2000, 10, 184

James Leggett, 35, 25, 500, 5, 50

John W. Westfall, 75, 980, 5025, 115, 321

Jos. B. Frederick 40, 290, 3300, 100, 400

John Gbourn (Osbourn), 200, 174, 4000, 100, 457

John W. Osbourn, -, -, -, -, 237

William Webb, 125, 680, 4330, 45, 448

Samuel Human, 100, 288, 2721, 50, 521

Enoch R. Hill, -, 170, 233, 20, 170

James Stuart, 30, 70, 250, 5, 81

John C. Nutter, 45, 75, 800, 5, 105

James A. Yates, 25, 175, 740, 10, 150

Wm. Modessett, 21, 104, 241, 5, 73

Nelson Richards, 18, 82, 200, 5, 12

James T. Smith, 70, 130, 1000, 5, 47

Jos. S. Hardman, 18, 64, 216, 12, 286

John H. Bell, 60, 400, 1000, 145, 323

Eli Cunningham, 8, 80, 220, 5, 35

John Modessett, 4, 46, 200, 5, 46

John Collins, 4, 46, 200, 5, 50

John Cunningham, 13, 73, 350, 5, 160

Mr. Amos Agt for Mr. Point, 30, 470, 1000, -, -

Nimrod South (Louth), 40, 60, 500, 5, 102

Nicholas H. Frederick, 36, 176, 1200, 60, 215

Saml. B. Frederick, 35, 112, 1200, 55, 200

Philip J. Frederick 50, 150, 1500, 20, 176

Jonathan Bessey, 20, 122, 600, 3, 64

Henry Fleming (Fling), 30, 100, 1200, 30, 443

Strother G. Goff, 50, 300, 1550, 10, 159

James Hardman, 50, 90, 1000, 20, 3345

P. S. Austin, 70, 1050, 6485, 60, 252

A. P. Hardman, 65, 60, 1000, 120, 377

P. J. Cunningham, 15, 253, 1300, 12, 15

Barnes Smith, 50, 151, 1000, 60, 392

Isaac Smith, 140, 342, 6495, 100, 762

Wm. H. Peters, ½, -, 300, -, 12

John B. Rogers, 100, 470, 3420, 103, 1014

Wm. Dilworth, 140, 211, 3159, 59, 351

Dennis Dye, 35, 172, 414, 16, 412

John R. Hostetter, -, -, -, 65, 442

Adam Laird, 30, 70, 300, 10, 137

Wm. L.(S.) Wilson, 20, 80, 300, 5, 77

Cyrus Lawson, 40, 60, 300, 15, 206

John Cain Jr., -, 23, 175, 35, 276

Alexr. Bickerstaff, 10, 153, 340, 5, 177

Andrew Hall, 95, 205, 4000, 90, 790

Jacob Moates, 175, 193, 3680, 40, 1129

John G. Hall, 200, 133, 7659, 18, 567

John Heaton, -, -, -, 80, 513

James Moates, 25, 137, 810, 3, 161

Christina Cokeley, 25, 70, 1200, -, -

Edmund Cokeley, 100, 500, 2000, -, -

Charles T. Lewis, 75, 181, 5120, 70, 473

Roane County, West Virginia
1860 Agricultural Census

The University of North Carolina at Chapel Hill filmed the 1860 agricultural census for Roane County from originals at the West Virginia State Archives under a grant from the National Science Foundation in 1963.

Columns 1, 2, 3, 4, 5, and 13 represent the following information on the census:
1. Name of Owner, Agent or Manager of Farm
2. Acres of Improved Land
3. Acres of Unimproved Land
4. Cash Value of the Farm
5. Value of Farming Implements and Machinery
13. Value of Livestock

James M. Payne, -, 126, 400, 65, 162
Wm. Gandell, 25, 300, 2000, 70, 200
H. F. Gibson, 20, 137, 1000, 10, 225
John Hughes, 30, 120, 800, 10, 250
W. E. Green, 25, 1200, 2500, 125, 300
John Good, 30, 394, 1000, 5, 320
Wm. Taylor, 40, 410, 1500, 75, 400
Jno. Cromwell, 35, 1265, 3000, 65, 400
John D. Campbell, 30, 1200, 1200, 38, 250
Elijah Campbell, 40, 70, 700, 24, 600
Jery Sword, 60, 240, 700, 8, 200
Thos. J. Jackson, 75, 325, 1200, 20, 500
A. B. Jackson, 25, 75, 400, 10, 125
James M. Armistead, 40, 487, 1500, 20, 600
Archibald Osborn, 30, 130, 2000, 20, 250
B. J. Taylor, 35, 280, 1000, 25, 100
Charles Osborn, 40, 210, 2000, 10, 200
Matthew Geary, 200, 1500, 10000, 100, 400
Isaac Osborn, 30, 255, 1500, 15, 100
John Smarr, 20, 280, 700, 10, 175

Michael R. Drake, 25, 25, 600, 5, 140
Daniel McGlothlin, 120, 150, 1400, 25, 500
Henry D. Sergent, 35, 160, 1000, 60, 500
Bradley Lowe, 70, 230, 1200, 6, 10
Wm. Hammack, 40, 125, 600, 10, 200
John H. Dougherty, 25, 93, 590, 5, 170
Madison Hively, 22, 138, 1400, 15, 100
John Hively, 125, 236, 3000, 100, 500
Alexr. Dougherty, 30, 131, 800, 6, 225
James R. Ryan, 20, 215, 800, 5, 200
Martin Hammack, 30, 102, 600, 5, 75
Zadoc Canterbury, 25, 350, 1100, 100, 150
Wm. Green, 20, 20, 200, 6, 80
Jacob Cummins, 75, 75, 800, 95, 200
Martin Summers, 25, 275, 900, 7, 150
David Thompson, 50, 50, 1500, 6, 210
James Moore, 22, -, 200, 6, 100
John Paxton, 60, 628, 600, 60, 350

Edward Lewis, 250, 950, 3500, 12, 800

John G. Ferrell, 40, 260, 1500, 10, 300

_____ Loontz (Looney), 70, 330, 3000, 75, 300

_. _. Ellis, 35, 300, 1500, 70, 230

John McLaughlin, 40, 60, 500, 15, 125

Aaron Noe, 35, 205, 1500, -, 100

James M. Moore, 30, 315, 1000, 12, 130

Geo. M. Tawney, 75, 520, 2000, 6, 300

J. H. & Isaac Moore, 100, 1000, 4000, 15, 800

Isac Drake, 30, 190, 750, 10, 100

D. D. Adkins, 130, 870, 5000, 50, 400

Thos. B. Cobb, 60, 340, 200, 5, 100

Wm. C. Moore, 35, 180, 100, 10, 150

James Patton, 25, 140, 200, 5, 150

Solomon Drake, 30, 250, 800, 20, 75

Wm. H. Justice, 45, 125, 800, 5, 180

Absalom Naylor, 40, 29, 400, 5, 30

Jacob Naylor, 30, 30, 200, 5, 30

Wm. Patton, 25, 175, 420, 5, 175

W. P. Knight, 30, 70, 300, 6, 175

James R. Knight, 100, 370, 2500, 10, 250

Wm. H. Sergent, 60, 840, 800, 5, 150

John King, 100, 300, 2000, 50, 400

Wm. Noe, 50, 498, 2000, 10, 250

David B. King, 70, 175, 2000, 6, 450

Elijah Rogers, 40, 680, 3000, 6, 200

Timothy O'Brien, 50, 350, 4500, 15, 125

James A. Cookman, 50, 250, 3000, 50, 250

John Smith, 70, 250, 1000, 130, 450

Solomon Salyers, 30, 210, 1000, 10, 90

Eli Harold, 40, 70, 175, 10, 75

Charles Drake, 40, 860, 3000, 7, 100

Archibald Collins, 40, 410, 1500, 9, 300

John White, 35, 240, 1000, 10, 180

Robt. F. Ogden, 35, 290, 1000, 5, 200

Aaron Hensley, 70, 200, 500, 10, 200

James Keer, 400, 1000, 15000, 100, 450

Samuel K. Smith, 30,175, 600, 5, 30

Wm. Ferrell, 50, 250, 700, 15, 200

John B. Stone, 75, 450, 2000, 200, 500

Wm. W. Noyes, 100, 140, 1500, 20, 300

Wm. Alfred, 200, -, 2000, 25, 300

Wm. Drake, 300, 560, 3000, 20, 250

Wm. Vinyard, 100, 300, 1000, 15, 85

John W. Kelly, 35, 22, 700, 5, 120

Abraham Looney, 100, 1800, 2500, 100, 550

Eva Ellis, 70, 85, 2000, 200, 550

George Hascue, 60, -, 300, 120, 200

John M. Jones, 140, -, 700, 150, 450

Geo. W. Martin, 40, 60, 500, 15, 200

Wilson Lowe, 40, -, 200, 10, 40

Samuel W. Gibson, 40, 160, 600, 10, 170

James N. Dougherty, 50, 70, 500, 5, 150

John Dolton, 45, 5, 250, 75, 250

A. J. Kelly, 200, 250, 2500, 100, 500

Christopher Hively, 30, 120, 1400, 15, 235

Daniel Looney, 100, 4100, 8000, 30, 400

David Boothe, 18, 182, 650, 5, 85

Wm. Tucker, 50, 375, 800, 75, 400

John W. Spencer, 100, 300, 600, 10, 350

Geo. King, 35, 175, 1000, 10, 400

Henry Chapman, 30, 70, 150, 10, 100

Kellis Chewning, 30, 570, 2800, 75, 350

Peter Looney, 100, 900, 3000, 15, 700

James Boggs, 120, 843, 2000, 25, 500

Ahab L. Collins, 12, 40, 120, 5, 100

B. S. Young, 51, 249, 1000, 6, 365

Wm. J. Batten, 40, 400, 1200, 6, 120

Ro. Ervin, 160, 2440, 9000, 175, 1000

Jeremiah Mace, 120, 440, 1500, 100, 650

Jacob Mace, 65, 425, 1000, 15, 200

Jefferson Tanner, 30, 115, 600, 10, 75

Wm. Heckert, 30, 100, 200, 6, 40

Daniel Bowers, 35, 100, 800, 35, 150

John Looney, 80, 90, 1800, 100, 70

Staple Handley, 35, -, 200, 9, 85

Hiram Cummins, 30, 120, 450, 5, 160

Wyatt J. Ferguson, 30, 90, 1000, 15, 200

Thomas Ferrell, 150, 1250, 6500, 75, 610

Bazle Compton, 40, 60, 500, 15, 100

Alex. Donalson, 35, 322, 357, 10, 225

Washington Cox, 30, 275, 450, 10, 175

A. Bowman, 51, 120, 1200, -, 275

A. G. Ingraham, 100, 267, 3000, 100, 300

Nathaniel Hardman, 30, 116, 800, 10, 80

M. B. Armstrong, 100, 790, 5000, 100, 300

Isaac McKown, 30, 10, 1000, 100, 200

Wm. R. Goff, 300, 4700, 10000, 75, 925

Jesse Tanner, 100, 75, 4000, 56, 300

Wm. Arnot, 30, 343, 750, 6, 90

Harrison Meadows, 40, 260, 800, 7, 110

Joshua A Meadows, 45, 188, 2000, 8, 158

Andrew J. Meadows, 12, 88, 500, 10, 130

Samuel H. Lowe, 40, 160, 800, 6, 200

Charles B. Lowe, 65, 397, 500, 60, 350

Joshua F. Arnot, 20, 230, 500, 9, 75

Samuel Sinnett, 35, 265, 1200, 6, 185

Zatt__ C. Ellis, 22, 182, 900, 5, 65

Thos. Mitchell, 15, 600, 1800, 75, 90

Andrew Coteral, 100, 106, 1500, 60, 425

A. D. Rohrbough, 10, 40, 150, 10, 70

Joseph A. Wright, 20, 80, 500, 7, 163

Wm. B. Whitzel, 40, 184, 1200, 5, 175

John Carpenter Sr., 20, 180, 1000, 5, 20

Wm. Allen, 35, 110, 900, 6, 40

H. T. Hughes, 60, 1000, 3000, 5, 300

Henry Payne, 75, 164, 2000, 100, 300

Wm. R. Wilson, 75, 100, 500, 6, 250

F. G. Frinnell, 30, 270, 1500, 7, 150

Jacob B. Morrison, 30, 270, 1500, 6, 300

Jos. B. Wolf, 60, 740, 2000, 75, 250

Henry Runnion Jr., 40, 460, 1800, 8, 175

C. F. Holswade, 100, 1200, 9000, 75, 400

Wm. Reynolds, 20, 80, 600, 5, 40

Nimrod Dawson, 17, 13, 300, 5, 40

William A. Davis, 75, 490, 2500, 6, 200

Jacob W. Riger, 26, 108, 1000, 5, 165

Jacob H. Bonnett, 30, 93, 300, 5, 300

Joseph Shouldis, 45, 59, 500, 6, 200

Stephen Hurt, 25, 71, 300, 7, 100

William Price, 25, 215, 600, 6, 250

W. P. Taylor, 12, 132, 500, 6, 80

Alex. Harper, 50, 90, 800, 45, 250

Bill Harlace, 161, 80, 100, 6, 200

Anthony Cook, 48, -, 100, 6, 75

Wm. F. Pell, 35, 20, 150, 6, 25

James Summers, 60, 137, 800, 10, 400

Wm. Harmmon, 30, 95, 500, 5, 150

John W. Price, 60, 340, 1600, 6, 200

John Hively, 50, 95, 1300, 40, 150

George Gandee, 50, 165, 1600, 8, 350

Asa Harper Sr., 80, 120, 3000, 150, 383

John D. Lynch, 50, 293, 1700, 95, 175

John Haynes, 25, 25, 200, 6, 150

John Greathouse, 100, 315, 4000, 5, 50

Elisha Collison, 50, 300, 700, 70, 175

Wm. Springston, 70, 170, 1000, 75, 250

_. J. Thomasson, 60, 397, 3000, 90, 150

_. L. W. Pool, 12, 216, 300, 5, 90

John Greathouse Jr., 30, 70, 600, 5, 65

James Riddle, 100, 40, 1500, 150, 500

_____ Greathouse, 14, -, 70, -, 150

John Miller, 37, 85, 610, 10, 250

James Wright, 75, 520, 1800, 75, 200

Wm. W. Thompson, 40, 80, 200, 30, 309

James A. Daniell, 100, 700, 5000, 60, 400

A. L. Vandal, 40, 660, 2000, 50, 400

Jacob S. Chambers, 30, 540, 1800, 10, 100

Marshal Dessne (Depue), 80, 97, 1500, 10, 300

Melton Webb, 40, 353, 1572, 8, 207

Wm. Fisher, 25, 210, 700, 5, 150

Leonard Epling, 30, 172, 1000, 8, 150

Harrison F. Harless, 30, 370, 1200, 3, 150

Benj. Hickel, 25, 875, 900, 10, 130

Delila Runnion, 25, 98, 12000, 6, 110

Samuel Hall, 25, 75, 500, 7, 100

Thos. Carpenter, 75, 635, 3500, 50, 350

Elias Summerfield, 25, 80, 600, 6, 300

Jonathan Smith, 35, 165, 1000, 6, 369

Joseph M. Cobb, 8, 42, 100, 3, 100

Benj. Summerfield, 25, 75, 500, 9, 155

Nicholas Smith, 20, 80, 400, 6, 150

Henry Nida, 25, -, 125, 5, 12

Peter Hammack, 25, 111, 200, 5, 200

St. Clair Hammack, 30, 353, 4500, 10, 200

Delana Vinyard, 65, 635, 3000, 15, 200

Presley Vinyard, 25, 175, 1500, 70, 120

John Vinyard, 40, 193, 1200, 6, 30

Fendal Salmon, 8, 60, 150, 5, 25

Robert Snodgrass, 45, 386, 2000, 25, 650

Francis Whited, 17, -, 85, 5, 200

James S. Gandee, 50, 168, 1500, 6, 200

Elijah Leforce, 25, 275, 1400, 9, 200

Elizabeth Bent, 50, 50, 500, 6, 100

James Lance, 20, 55, 700, 6, 125

George Lance, 50, 50, 800, 75, 250

Jesse West, 30, 200, 850, 6, 250

Mark Hersman, 25, 200, 1500, 5, 200

Marsh Cunningham, 25, 75, 800, 6, 75

Noah Lawrence, 76, 84, 800, 5, 150

Jesse Coteral, 45, 55, 500, 6, 100

Christopher Tompkins, 40, 60, 1000, 5, 195

Dempsey P. Flesher, 100, 120, 2200, 50, 200

John W. Stewart, 30, 200, 1000, 20, 100

Jacob B. Smith, 35, 365, 800, 30, 100

Alex. S. Boord, 50, 250, 1500, 50, 300

T. A. Roberts, 20, -, 1000, 150, 500

Alfred Cain, 80, 70, 3000, 100, 600

Charles Roach, 40, 110, 1500, 25, 220

Silas B. Seaman, 200, 1600, 7000, 100, 500

Wm. K. Boord, 12, 13, 400, 8, 170

Matthew McLaughlin, 20, 43, 250, 10, 100

Wm. H. Greathouse, 100, 300, 2000, 50, 350

Wm. Boggs, 25, 131, 1248, 10, 85

Zilla Ward, 100, 500, 3000, 75, 450

Benniah Depue, 135, 851, 3544, 75, 700

John Amrick, 20, 180, 1000, 75, 100

Leonard Simmons, 150, 598, 2244, 100, 900

Samuel Miller, 100, 290, 1500, 100, 400

John R. Boggs, 60, 337, 1190, 80, 300

Jas. W. Greathouse, 40, 55, 285, 10, 250

Geo. W. Feather, 25, -, 100, 40, 100

Abel Strader, 50, 225, 1500, 20, 250

James Simmons, 120, 130, 140000, 95, 300

Alan D. Hadan, 80, 239, 2000, 75, 200

M. D. N. Boggs, 125, 332, 4000, 100, 600

Wm. B. Vandal, 50, 50, 1200, 75, 200

Josiah Stutler, 35, 265, 900, 30, 300

Thomas D. Goff, 20, 80, 300, 10, 100

Christopher Stutler, 35, 265, 1000, 20, 200

H. B. Butcher, 40, 93, 700, 6, 75

Elizabeth Watts, 70, 180, 1000, 90, 400

Thos. H. Cain, 50, 30, 1000, 70, 350

Samuel Wyatt, 50, 170, 1000, 70, 350

Martin Sims, 180, 122, 4000, 75, 250

James Cummins, 50, 220, 900, 50, 178

James D. Seaman, 30, 70, 500, 6, 160

Jesse B. Knopp, 80, 470, 2200, 100, 350

Ira S. Chenoweth, 50, 350, 1800, 100, 500

John M. Smith, 20, 300, 1500, 8,185

James Welch, 75, 253, 2000, 27, 120

Francis Snyder, 70, -, 100, 5, 80

John Stalnaker, 60, 340, 800, 100, 175

John H. Rader (Reeder), 70, 450, 1300, 60, 250

David A. Lattimore, 100, 30, 3900, 5, 375

Samuel Hall, 80, 320, 200, 100, 250

Flavis Parsons, 75, 200, 825, 75, 250

Wm. M. Crookshanks, 20, 380, 1200, 8, 200

M. A. Seaman, 40, -, 200, 8, 250

David Seaman, 30, -, 150, 6, 120

Joseph Stuart, 25, -, 150, 10, 100

John Flesher, 90, 400, 1245, 30, 350

Wm. Stuart, 100, 220, 1200, 100, 300

Hiram Chancey (Chaucey), 100, 190, 1590, 100, 250

Andrew B. Chancey (Chaucey), 15, 80, 200, 8, 150

Roswell R. Chancey (Chaucey), 20, 70, 2000, 10, 150

Calvary Chaucey (Chancey), 18, 82, 300, 5, 120

Wm. Roach, 200, 500, 8000, 65, 400

Samuel Rhodes, 60, 120, 1800, 5, 175

George S. Goff, 100, 60, 400, 100, 300

Henry W. Smith, 30, 170, 400, 20, 175

Samuel Argabrite, 25, 55, 300, 5, 200

Elijah Beneditt, 75, 1445, 1400, 100, 300

James M. Sergent, 35, 265, 900, 23, 75

Bailey Clevenger 30, 625, 2640, 100, 250

Hugh Kyger, 140, 1860, 10000, 100, 700

James C. Haid, 12, -, 72, -, 20

Edmond Budget, 50, 700, 1200, 200, 250

Dempsey Flesher, 75, 187, 500, 5, 45

John C. Lester, 25, 197, 1000, 50, 180

R. M. Kyger, 80, 320, 3000, 100, 250

Jacob C. Smith, 75, 200, 1500, 75, 400

Elijah Callow, 40, 120, 1200, 100, 300

Geo. W. Callow, 35, 135, 900, 75, 300

John G. Goff, 40, 90, 1500, 10, 95

Joseph Boon, 100, 542, 5000, 10, 175

Wm. J. Riddle, 15, 65, 600, 3, -

Stewart S. Ingraham, 12, 123, 500, 6, 35

Benj. Riddle, 100, 300, 4000, 100, 650

John D. Smith, 13, 2, 100, 5, 40

Pleasant H. Thomasson, 100, 360, 4000, 75, 350

John S. McCauley, 40, 260, 1500, 5, 75

John Hoff, 40, 160, 1500, 15, 200

James E. Burdett, 50, 950, 5000, 50, 400

Thomas Hardman, 50, 200, 1500, 75, 250

Geo. M. Hardman, 25, 25, 250, 5, 35

Henry Nelson, 175, 525, 5500, 50, 530

John J. P. Armstrong, 100, 790, 2000, 60, 100

Kellis M. Argabrite, 30, 45, 700, 5, 125

A. H. Parsons, 25, 135, 700, 5, 50

Wm. Hardman, 45, 48, 1000, 65, 375

Wm. W. Curtis, 50, 323, 1200, 20, 255

Geo. A. Argabrite, 30, 170, 400, 5, 200

James A. Davis, 20, 80, 600, 5, 25

Jonathan Rollins, 100, 700, 800, 70, 287

Wm. R. Randolph, 30, 220, 500, 10, 300

Wm. D. Kelly, 40, 190, 1200, 15, 75

John M. Ranes, 60, 165, 1000, 15, 250

Mathias Rhodes, 75, 130, 1000, 50, 200

David Taylor, 25, 75, 600, 10, 40

Wm. Riley, 40, 1460, 1500, 15, 300

Joseph Rhodes, 65, 125, 190, 10, 175

Geo. W. Riley, 60, 1290, 5000, 15, 350

Abram Ranes, 50, 250, 600, 60, 250

Stephen Carpenter, 60, 650, 700, 25, 85

James Lowe, 25, 238, 800, 10, 60

Joshua Hammon, 35, 127, 1500, 6, 150

Joseph Presley, 75, 85, 800, 6, 300

John Smith, 40, 160, 250, 25, 250

Joel Cunningham, 20, 200, 3000, 6, 250

Josiah Hughes, 60, 44, 1040, 50, 300

Jeremiah Coreby, 45, 306, 1200, 40, 200

Jacob A. Cuislip, 40, 358, 1500, 63, 200

Michael Ronnine, 25, 9, 300, 6, 175

Hudson Dearman, 30, 170, 600, 5, 140

Washington Casto, 40, 160, 600, 10, 250

Eli Perkins, 50, 350, 2000, 50, 150

Simon A. Davis, 25, 36, 600, 40, 330
Lyle Paxton, 35, 400, 435, 15, 200
Geo. H. Cunningham, 35, 115, 1500, 35, 250
Anthony L. Hoff, 16, 80, 480, 6, 73
George Lance, 45, 75, 1000, 125, 400
John McCoy, 100, 235, 2000, 40, 400
Lemuel Cuislip, 80, 170, 1500, 15, 450
Thomas Miller, 40, 60, 800, 40, 200
Jeremiah Miller, 50, 50, 500, 20, 150
Amos Miller, 30, 81, 600, 6, 65
Archibald Kelly, 60, 180, 1000, 10, 200
George O'Neal, 30, 120, 525, 10, 120
Allen Cuislip, 45, 65, 1000, 6, 200
Joseph F. Engle, 20, 46, 400, 5, 160
Richard Greathouse, 30, 70, 500, 5, 120
Geo. Hanger (Hauger), 50, 104, 600, 35, 80

Jonathan Simmons, 23, 177, 900, 5, 150
Stephen Freeland, 35, 265, 600, 6, 175
Salathiel S. Harman, 45, 30, 600, 5, 150
James A. Harper, 140, 388, 1000, 10, 150
Lewis A. Phillips, 30, 70, 200, 25, 200
Armistead Harper, 130, 149, 800, 25, 275
Eperson Harper, 75, -, 300, 15, 350
Jolson Givens, 40, 217, 1000, 15, 250
John L. Cook, 25, 125, 175, 5, 250
John Shaver, 15, 64, 250, 5, 100
Jacob Shaver, 20, 110, 200, 5, 135
Wm. R. Clarkson, 40, 147, 1000, 35, 140
Levi H. Hunt, 15, 60, 300, 5, 135
Geo. W. Fields, 60, 419, 6000, 25, 400
John Harper Sr., 70, 90, 1500, 18, 210

Taylor County, West Virginia
1860 Agricultural Census

The University of North Carolina at Chapel Hill filmed the 1860 agricultural census for Taylor County from originals at the West Virginia State Archives under a grant from the National Science Foundation in 1963.

Columns 1, 2, 3, 4, 5, and 13 represent the following information on the census:
1. Name of Owner, Agent or Manager of Farm
2. Acres of Improved Land
3. Acres of Unimproved Land
4. Cash Value of the Farm
5. Value of Farming Implements and Machinery
13. Value of Livestock

Adolphus Armstrong, 120, 480, 9800, -, 150
James W. Batson, 150, 50,800, 25, 115
Harman Sinsel, 15 ½, -, 1000, 20, 65
M. H. Johnson, 600, 400, 12000, 300, 1500
E. J. Armstrong, 400, 175, 13000, 100, 1800
J. S. Burdett, 100, 50, 4000, 20, 350
Abraham Smith, 900, 300, 24000, 400, 3500
William Wheeler, 100, 200, 3000, 50, 340
Frederick Burdett, 100, -, 2500, 10, 170
George Brown, 109, 25, 2500, 100, 500
C. E. Reynolds, 500, 400, 10000, 200, 1500
Thomas Gawthrop, 75, 27, 1700, 160, 225
S. W. Tolls, 50, 50, 1500, 100, 380
Levi Barker, 80, 220, 3500, 50, 200
David Coplin, 45, 9, 1000, 100, 280
Moses Husted, 190, 210, 8000, 75, 495
Thomas Cather, 400, 160, 17500, 100, 2000

P. F. McDonald, 60, 32, 2000, 50, 500
Joseph Taylor, 100, 100, 5000, 100, 960
John McDonald, 40, 230, 3100, 125, 760
P. W. Knight, 80, 35, 2875, 150, 500
Richard Hosking, 200, 100, 7500, 50, 270
James H. Fletcher manager, 70, 54, 2500, 125, 250
Jos. Finley (renter), 50, 55, 1200, 50, 150
B. A. Coplin, 100, 62, 3000, 50, 290
Zackquille Cockran, 100, 60, 3400, 100, 730
Thomas Yates, 40, 34, 1100, 80, 200
Jesse Husted, 50, 4, 1280, 100, 300
G. W. Husted, 60, 43, 1400, 50, 270
William Riffee, 30, 13, 400, 25, 200
Narcissa Singleton, 50, 73, 960, 50, 250
B. F. Payne, 200, 76, 5000, 100, 1220
George Corbin, 30, 20, 1000, 150, 280
Abner Abbott, 60, 40, 2500, 70, 378
James Ryan, 70, 50, 3000, 150, 455
A. Freeman, 140, 100, 3000, 130, 375

W. Freeman, 20, 3, 450, 50, 230

G. T. Martin, 1000, 500, 49000, 300, 5050

Gabriel Smith, 40, 36, 760, 100, 300

R. P. Nixon, 70, 46, 1500, 50, 292

Jos. H. Lambert, 50, 25, 700, 20, 123

Melinda Laughlin, 40, 10, 800, 25, 198

George Freeman, 40, 30, 700, 30, 135

Greenberry Wilson, 40, 60, 1200, 125, 315

Jacob Smith, 40, 35, 800, 65, 280

John Rogers, 100, 47, 3500, 110, 225

Mary A. Ryan, 100, 20, 2500, 120, 320

John R. Elder, 70, 55, 2200, 120, 320

James E. Riley, 65, 18, 1500, 65, 315

Jefferson Keener, 80, 60, 2550, 60, 390

Elias B. Glenn, 50, 23, 1300, 40, 186

William Johnson, 35, 52 ½, 1500, 25, 240

A. S. Keener, 50, 49, 1250, 75, 290

Wilson Brown, 200, 100, 6000, 90, 1009

Adam Zambro, 70, 127, 2600, 75, 315

Hiram Linn, 250, 263, 6500, 75, 734

John McWilliams (Renter), 70, 30, 1500, 15, 74

William Mallones, 280, 58, 13000, 90, 1105

Robert Starke, 250, 195, 6000, 80, 340

A. McElfresh, 60, 40, 2000, 65, 227

Andrew Hertzog, 80, 175, 3500, 60, 338

M. S. Corbin, 50, 30, 1000, 50, 319

William Smith, 100, 40, 2900, 65, 325

William Gray, 40, 10, 500, 30, 275

John Lawler, 15, 36, 600, 25, 155

James P. Carrey, 66, 65, 2300, 50, 445

Drucilla Tucker, 66, 14, 1600, 35, 60

James Clelland, 100, 181, 4820, 85, 784

Melinda Reed, 70, 20, 1800, 20, 172

Sarah Hoult, 100, 50, 3000, 45, 205

Thomas H. Jones, 100, 128, 4000, 200, 856

Joseph B. Jones, 40, 45, 1700, 10, 222

John S. Tucker, 75, 85, 2800, 100, 340

T. F. Harr (Starr), 175, 145, 6400, 80, 420

Isaac B. Carder, 70, 85, 2325, 40, 290

C. S. Whittaker, 100, 46, 2600, 65, 325

John A. Gawthrop, 200, 200, 8000, 25, 202

Enoch Dunham renter of Joseph Carr, 70, 25, 1200, 40, 250

David Dunham, 50, 76, 1386, 40, 220

Francis Coplin, 260, 208, 7910, 100, 1193

Alex. Williamson, 50, 35, 1700, 75, 450

James M. Starr, 40, 55, 1710, 25, 185

J. F. Carrey, 138, 138, 4000, 100, 423

William Scranage, 100, 200, 4325, 60, 400

Benjamin Currey (Carrey), 40, 16, 840, 25, 420

George Coffman, 60, 42, 1125, 30, 120

John Henderson, 40, 60, 1000, 20, 145

Bailey Knight, 40, 40, 1870, 15, 270

S. B. Frum, 16, 34, 750, 10, 175

Wilbid Watkins, 10, 38, 850, 5, 135

Willis Lawler, 120, 180, 6000, 125, 640

John Sinclair, 150, 50, 3000, 75, 530

Aquilla Martin, 100, 137, 4700, 85, 560

E. B. Smith, 60, 116, 1760, 25, 258

James C. Wiseman, 20, 60, 640, 20, 67

Marshal Wiseman, 50, 45, 1241, 25, 290

Bennet Wheeler, 200, 300, 7000, 100, 540

J. B. Lawler, 40, 46, 860, 15, 250

A. Williamson, 70, 52 ½, 1220, 10, 355

James D. Scranage, 75, 125, 3000, 25, 320

Silas Giles, 40, 27 ½, 1500, 15, 290

Reuben Bennet, 90, 46, 2720, 20, 340

Alex. Scranage, 180, 10, 4000, 205, 825

Richard Harr (Hass), 180, 130, 4600, 125, 867

William Linn, 70, 330, 3200, 100, 450

John Peters, 50, 32, 1215, 25, 200

James Sinclair, 100, 60, 5280, 115, 500

Benjamin McDonald, 80, 10, 5000, 125, 750

Robert Lowe, 275, 75, 8740, 120, 1570

Edwin McGee, 20, 500, 2000, 30, 170

Aurelins Goff, 270, 100, 9250, 100, 1705

Jonathan Currey, 9, 6, 180, 10, 30

William Brown, 30, 20,750, 5, 40

Robert Johnson, 30, 23, 1800, 30, 300

Caleb B. Stanley, 50, 53, 1500, 25, 382

Robert Brown, 100, 190, 2900, 50, 275

Samuel Brown, 75, 27, 2000, 30, 230

George Wiseman, 70, 30, 1000, 80, 360

William Henderson, 45, 40,680, 25, 165

Edward Henderson, 40, 44, 672, 18, 225

Isaac Reese, 75, 25, 2000, 25, 240

John Musgrave, 175, 362, 5552, 100, 505

Nimrod A. Lake, 60, 40, 1000, 20, 200

Thomas Henderson, 40, 23, 588, 15, 180

F. B. Henderson, 120, 60, 2000, 25, 295

W. G. Henderson, 100, 137, 2284, 95, 306

S. W. Henderson, 45, 40, 850, 65, 340

Gabel Reese, 100, 40, 1000, 20, 157

Isaac Vangilder, 35, 73, 900, 15, 154

David Summers, 150, 255, 3767, 10, 400

John Miller, 70, 130, 1500, 100, 300

Patrick Bradley, 60, 40, 700, 10, 125

A. J. Corothers, 75, 175, 3000, 80, 360

William Corothers, 85, 88, 3000, 10, 476

Henry Roderick, 65, 152, 2170, 15, 200

John Rogers, 50, 50, 1000, 10, 175

Samuel Corothers, 320, 205, 8000, 100, 1500

A. Reed, 30, 70, 1000, 20, 180

John Shanholtz, 40, 28, 406, 10, 150

W.R. Hunt, 40, 28, 404, 12, 160

Robert Luzader, 30, 43, 400, 10, 180

David Roderick, 80, 71, 1500, 50, 250

Samuel Watkins, 60, 73, 1600, 25, 250

Shelton Ford, 150, 150, 4000, 40, 460

George Shehan, 40, 50, 800, 20, 307

E. S. Shackelford, 30, 70, 1000, 15, 300

James Rogers, 10, 58, 500, 10, 150

Aaron Luzader, 50, 50, 884, 15, 235

Robert Reese, 40, 60, 800, 10, 100

Robert Boyd, 80, 220, 4000, 75, 592
John Rinker, 20, 15, 720, 120, 278
David R. Gary, 50, 27, 200, 75, 400
Jesse W. Carder, 25, 11, 600, 20, 110
Moses Hardinger, 75, 25, 1150, 25, 434
Jonathan Poe, 75, 145, 2200, 75, 229
William Keener, 100, 121, 1105, 50, 605
Thomas Devers, 50, 115, 1550, 75, 200
S. W. Poe, 374, 374, 7480, 25, 1130
E. Miller, 39, 43, 800, 10, 100
Alex. Courtney, 60, 40, 1500, 15, 356
M. C. Murphey, 100, 4, 3000, 75, 580
J. A. Poe, 175, 97, 4000, 25, 772
Saml. Philips, 70, 30, 700, 10, 250
Perry Moran, 50, 50, 600, 15, 250
F. Matthews, 32, 145, 1720, 15, 225
James Poe, 100, 60, 355, 20, 506
Benj. Matthews, 50, 127, 1720, 15, 295
William Rogers, 85, 81, 6000, 65, 450
James Haymond, 100, 135, 2500, 130, 1000
Joseph Williams, 130, 120, 3500, 80, 290
James Austin, 50, 50, 1000, 35, 374
O. P. Austin, 12, 3, 300, 10, 100
W. Rogers Jr., 50, 50, 700, 15, 250
N. Osborn, 100, 69, 2000, 100, 450
William Boyce, 50, 69, 1000, 25, 250
James Jacobs, 55, 20, 750, 30, 425
John Haymond, 200, 300, 5000, 100, 450
W. H. Grimes, 40, 85, 1000, 10, 250
Wm. Howard, 25, 46, 568, 12, 150
John Haymond Jr., 50, 100, 2000, 35, 250
Francis Poe, 150, 200, 3000, 60, 500
John Keener, 50, 87, 15000, 100, 463

Samuel Keener, 80, 74, 1540, 80, 860
John K. Knotts, 160, 100, 5000, 100, 660
W. B. Poe, 35, 50, 500, 25, 300
John Wood, 60, 56, 1100, 15, 200
Jacob Poe, 75, 45, 200, 150, 375
William Blue, 70, 38, 2160, 150, 350
Enoch Current, 70, 160, 2300, 20, 100
Abraham Wilson, 210, 90, 6000, 230, 1090
J. G. Means, 80, 120, 300, 85, 260
Abraham Luzader, 30, 21, 500, 10, 130
Jas. P. Menear, 40, 180, 2200, 15, 100
William Menear, 50, 33, 1000, 10, 150
John D. Keener, 70, 166, 1500, 20, 300
William Snider, 50, 58, 800, 30, 400
James Current, 75, 60, 1500, 50, 250
James Gatlin, 35, 65, 1000, 10, 80
Francis Warthew, 80, 70, 1500, 25, 300
A. W. Moore, 100, 169, 2000, 20, 500
Calder Haymond, 70, 169, 4000, 100, 600
Isaac Thomas, 180, 65, 1500, 75, 400
James Thomas, 35, 35, 700, 20, 250
A. Rightmire, 80, 200, 3000, 120, 255
J. Fawley, 120, 75, 2000, 60, 225
James Knotts, 300, 253, 5530, 60, 570
Jacob McCartney, 80, 68, 1000, 75, 275
G. McCartney & 3 others, 70, 58, 1000, 25, 200
Andrew Nose, 50, 51, 800, 50, 250
James Ford, 40, 65, 1000, 25, 225
J. M. & J. A. Wilson, 150, 80, 6000, 200, 897

Daniel Medsker, 50, 150, 2000, 25, 175

John Miller, 130, 70, 4000, 50, 550

Absalom Knotts, 75, 65, 1400, 25, 380

Jacob Means, 400, 300, 1000, 300, 1300

W. Ludwick, 200, 220, 5000, 100, 650

S. McDaniel, 40, 60, 1000, 20, 270

S. W. McDaniel, 150, 150, 3000, 50, 600

John W. Blue, 175, 125, 7500, 75, 400

S. B. Keener, 50, 112, 2430, 20, 200

A. M. Hesser lessor of G. L. Bitzer, 140, 232, 6000, 20, 200

George Smell, 100, 260, 5400, 100, 600

W. A. Means, 120, 100, 4000, 100, 1350

Jacob Shroyer, 150, 50, 3000, 100, 1000

Hugh Evans, 140, 160, 6000, 150, 1322

Jacob Hull, 75, 132, 4000, 125, 505

A. G. Hull, 70, 50, 1800, 20, 380

James Poe, 60, 21, 800, 30, 250

John A. Gusman, 140, 350, 2800, 100, 300

John Hull, 65, 30, 1500, 25, 150

Geo. Hull, 75, 95, 2000, 35, 454

J. C. Woodyard, 75, 329, 2020, 30, 540

John Lewellen, 150, 144, 3000, 100, 600

Jacob Rosier, 75, 50, 1000, 25, 250

John Rosier, 100, 25, 2000, 30, 445

Matthew Luzader, 100, 180, 3896, 100, 450

Moses Luzader, 70, 30, 1500, 120, 386

Lanty Ford, 120, 202, 6000, 100, 610

W. G. Poe, 250, 226, 4760, 125, 600

John Carr, 80, 45, 2500, 100, 250

William Means, 75, 163, 3000, 100, 750

Fleming Jones, 75, 125, 2000, 50, 500

Samuel Jones, 120, 20, 2000, 50, 450

A. B. Gawthrop, 300, 100, 6000, 100, 950

James Bartlett, 220, 269, 6845, 100, 1000

Daniel Woodyard, 200, 125, 600, 200, 1200

Noah Warder, 90, 26, 3000, 120, 590

Nathan Rector, 40, 60, 1800, 50, 350

J. M. Thayer, 75, 34, 4000, 300, 1000

John Payne & 4 others, 100, 120, 3450, 60, 520

David Elliott, 300, 280, 14000, 300, 1200

Saml. Waller, 20, 21, 800, 20, 35

N. E. Bartlett, 100, 70, 2800, 50, 500

William Findley, 200, 356, 6882, 125, 1340

John Sinsel, 150, 20, 5000, 50, 755

W. H. Shields, 100, 300, 5000, 200, 390

J. H. Barnes, 200, 98, 9000, 150, 600

Samuel West, 100, 50, 3500, 60, 450

James H. Sinsel, 50, 70, 3000, 30, 350

William Sinsel, 150, 53, 4284, 40, 325

Hary A. Newlon & 8 others, 80, 31, 4000, 60, 400

Alener Yates (Gates), 70, 46, 2600, 20, 250

Sarah Frest & 11 others, 75, 26, 2000, 50, 250

Stephen Gilbert, 70, 15, 2500, 30, 425

Mary Dillon, 100, 75, 2000, 25, 250

W. P. Bartlett, 50, 51, 2000, 15, 500

John Sinclair, 75, 45, 2440, 10, 380

J. W. Bartlett, 80, 80, 3000, 20, 650

T. T. Bartlett, 80, 40, 1850, 15, 100

M. B. Sinsel, 100, 127, 5000, 20, 190

L. D. Davis, 50, 20, 1750, 15, 300

Jas. K. Bartlett, 50, 69, 2150, 20, 450

James H. Bartlett, 25, 20, 1000, 10, 200

M. B. Lake, 20, 40, 500, 10, 180

Patrick Fleming Jr., 90, 87 ½, 3550, 50, 750

M. S. Fleming, 175, 150, 5000, 50, 908

Silas H. West, 110, 38, 4400, 10, 354

Thomas Newlon, 90, 35, 3125, 40, 370

Chapman Husted, 50, 170, 300, 30, 570

Thomas Selvey (Selney), 89, 36, 2900, 100, 800

William Newlon, 100, 200, 6000, 120, 900

Thomas J. Holbert, 60, 32, 2000, 40, 350

Alex Davidson, 80, 30, 2050, 30, 250

Alfred Rector, 124, 12, 3600, 125, 32

William Sinclair, 22, 59, 840, 20, 350

W. A. Lake, 90, 210, 3500, 60, 375

Benjamin Knight, 100, 41, 3525, 50, 500

Isaac Sinclair, 25, 41, 1000, 20, 250

Joseph Warder, 70, 87, 200, 40, 350

William Warder, 60, 36, 1500, 30, 250

William Cather, 115, 40, 3875, 75, 1000

Henry Warder, 130, 110, 7200, 30, 300

Reuben Davisson Jr., 450, 960, 17000, 75, 1725

B. N. Mason, 150, 150, 4000, 35, 660

Obediah Ford, 85, 35, 2400, 100, 640

William Bartlett, 80, 25, 2100, 75, 555

W. R.Coplin, 80, 20, 3500, 120, 560

John W. Cole, 300, 300, 9000, 100, 6335

John T. Cather, 50, 135, 3700, 25, 335

Daniel McVicker, 60, 34, 1000, 20, 200

Isaac Carder, 250, 350, 6000, 100, 500

John Curry, 196, 91, 5700, 50, 800

Geo. W. Robinson, 80, 70, 4100, 100, 1000

Elijah Sinsel, 320, 20, 8800, 120, 1400

Martin Yates (Gates), 100, 200, 6000, 50, 800

L. D. Davis, 75, 35, 2775, 20, 350

Silas Utterback, 215, 40, 7650, 150, 800

James Selvey, 80, 120, 4000, 150, 1100

John Whitehair, 40, 65, 1050, 20, 150

W. S. Bartlett, 60, 20, 880, 20, 360

Joseph West, 130, 120, 4000, 100, 800

Robert Rogers, 80, 141, 3000, 50, 250

Geo. Sharp, 250, 70, 7620, 150, 1400

Fielding Riley, 75, 400, 2100, 50, 350

Moses Gratehouse, 200, 3, 8000, 60, 850

Jasper Cather, 12, 156, 2000, 30, 500

Emanuel B. Smith, 80, 50, 5000, 120, 1600

J. C. Fleming, 225, 50, 7000, 100, 1000

L. E. Davisson, 85, 95, 3600, 30, 650

James Bailey, 125, 199, 7000, 120, 1050

Thomas Bailey, 400, 200, 10000, 250, 2200

Mary Bailey, 100, 40, 3500, 100, 800

John Robinson, 100, 75, 5250, 150, 830

James Monrow, 100, 100, 4000, 50, 700

Thomas Hawkins, 48, 2, 1000, 50, 800

Nimrod Bashard, 100, 50, 3000, 100, 750

William Bailey, 130, 160, 5000, 120, 700

Elijah Powell, 150, 75, 5000, 150, 1200

J. H. Cather, 350, 150, 12000, 180, 1546

Bailey Knight, 100, 105, 4100, 100, 600

Tucker County, West Virginia
1860 Agricultural Census

The University of North Carolina at Chapel Hill filmed the 1860 agricultural census for Tucker County from originals at the West Virginia State Archives under a grant from the National Science Foundation in 1963.

Columns 1, 2, 3, 4, 5, and 13 represent the following information on the census:
1. Name of Owner, Agent or Manager of Farm
2. Acres of Improved Land
3. Acres of Unimproved Land
4. Cash Value of the Farm
5. Value of Farming Implements and Machinery
13. Value of Livestock

Nathaniel Nester, 25, 75, 600, 15, 215
Elias Nester, 251, 65, 600, 15, 115
Geo. M. Nester, 22, 80, 600, 10, 110
John Shafer, 45, 195, 1800, 5, 400
James D. Nester, 40, 300, 1200, 15, 200
Samuel Shafer, 20, 200, 500, 5, 150
Eltin Hovater, 12, 188, 300, 3, 75
George Shahan, 61, 94, 300, 5, 12
George Parsons, 4, 96, 100, 5, 100
Elizabeth Manier, 20, 32, 600, 20, 125
Theodore Lipscomb, 18, 10, 400, 10, 150
John O. Robinson, 50, 86, 1000, 5, 200
James C. A. Goff, 30, 56, 800, 5, 40
Wm. Burn, 8, 130, 300, 5, 250
Amaca Goff, 35, 145, 820, 25, 450
John C. Goff, 50, 90, 1000, 9, 57
Adam H. Bacome, 90, 200, 2000, 20, 600
Aaron Loughey Sr., 50, 170, 1500, 20, 250
Aaron J. Loughey, 15, 335, 500, 5, 100
Jacob Spingly, 5, 140, 300, 5, 30
Novil T. Croston, 28, 65, 1000, 2, 500

John J. Cline, 2, 100, 200, -, 40
William P. Hobb, 60, 470, 1730, 50, 200
Bazel Moats, 60, 540, 5000, 5, 50
Levi Lipscomb Jr., 8, 32, 100, 3, 140
Levi Lipscomb Sr., 50, 100, 1000, 10, 175
Isaac S. James, 67, 250, 800, 15, 200
John H. James, 85, -, 1000, 100, 150
Daniel C. Adams, 75, 115, 2000, 10, 450
James W. Miller, 50, 72, 1000, 10, 300
John J. Adams, -, 250, 500, 10, 125
Andrew D. Moore, 100, 307, 1500, 15, 450
John White Jr., 30, 460, 1000, 15, 350
Jacob Dumire, 50, 140, 1000, 50, 340
Adam Dumire, 16, 84, 300, 15, 150
John W. Dumire, 30, 70, 500, 5, 95
John P. Gray, 60, 140, 500, 50, 500
John Nevil, 40, 150, 900, 10, 200
Robt. Jones, 12, 60, 500, 2, 30
Wm. T. White, 50, 100, 1000, 30, 300
John Whitison, 140, 260, 5000, 30, 400

Daniel Dumire, 60, 110, 1000, 10, 200

Adam White, 26, 760, 5000, 20, 350

Ephraim H. James, 35, 327, 1000, 5, 250

James Lipscomb, 9, 146, 300, 3, 100

Stephen Dumire, 20, 127, 600, 5, 200

Rhinehad Dumire, 50, 100, 1200, 15, 175

Joseph Kephart, 20, 80, 300, 5, 125

Patrick McCinny, 45, 100, 800, 5, 150

Frederick Shafer, 20, 136, 500, 40, 150

Isaac Woolf, 40, 158, 300, 5, 250

Joseph Hershline, 18, 134, 300, 3, 80

Henry Shrader, 35, 3, 560, 3, 200

Jno. M. Porter, 20, 400, 400, 2, 75

George Specard, 15, 100, 300, 2, 150

George Firet (Fint), 25, 200, 500, 2, 100

Henry V. Ark, 20, 18, 100, 3, 90

Samuel Rudolph, 80, 375, 2580, 100, 375

Jacob B. Wootring, 25, 800, 500, 6, 175

Frederick Dumire, 16, 16, 100, 5, 150

Daniel L. Dumire, 20, 69, 500, 5, 100

Robt. Knotts, 55, 95, 500, 5, 300

Martin L. Knotts, 12, -, 100, 5, 100

David Close, 50, -, 500, 5, 200

Wm. Thompson, 20, -, 200, -, 195

Stephen Losh, 40, 32, 500, 1, 70

James Evans, 5, -, 25, 1, 50

Wm. Losh, 70, 40, 1000, 20, 250

Nicholas M. Parsons, 250, 700, 8000, 250, 1500

Rufus Maxwell, 35, 123, 1100, 10, 160

Arnold Boremfield, 200, 1500, 5000, 300, 4500

Dr. Solomon Parsons, 100, 273, 400, 100, 300

Geo. B. Lee, 150, 450, 2500, 10, 500

Andrew Harsh, 30, 149, 800, 20, 250

Wm. Miller, 40, 80, 1000, 10, 230

Johnson Tolbott, 100, 260, 2500, 20, 500

Wm. Marsh, 80, 220, 2000, 20, 350

Francis D. Tolbott, 75, 125, 2000, 50, 200

Enoch Manier (Mauier), 150, 395, 6000, 50, 400

Jacob Nester, 25, 75, 500, 10, 100

John Anvil, 42, 35, 800, 16, 300

Andrew Pifer, 40, 90, 1000, 40, 300

John Toakman, 30, 17, 800, 20, 300

Wash. A. Long, 25, 100, 400, 10, 150

James Long, 100, 512, 3000, 15, 200

Jonathan Murphy, 25, 75, 500, 10, 125

Richard Mitchell, 30, 63, 400, 5, 150

Peter J. H. Linzey, -, 130, 150, 5, 150

Peter Shafer, 35, 295, 600, 50, 225

Jacob Shafer, 30, 67, 800, 10, 250

Israel Phillips, 60, 191, 633, 25, 350

John W. Phillips, 40, 147, 600, 10, 250

John W. Cross, 20, 141, 330, 10, 225

Aaron Phillips, 30, 100, 500, 5, 175

John Fitzwater, 50, 192, 600, 5, 123

Robt. Phillips, 25, 75, 500, 5, 40

David Keller, 30, 50, 200, 5, 5

Wm. Valentine, 60, 187, 1200, 25, 250

Enoch Phillips, 50, 120, 800, 10, 125

Moses Phillips, 12, 138, 400, 10, 150

Samuel Kaylor, 12, 136, 350, 10, 120

Joseph Myers, 15, 125, 200, 3, 100

John Kaylor, 45, 187, 1000, 50, 400

Wm. J. Harper, 150, 3850, 10000, 50, -

Adam Harper, 500, 1000, 5000, 50, 1000

Wm. Ervin, 300, 1600, 10000, 100, 2800

Saml. W. Bowman, 125, 156, 2000, 100, 300

Jas. W. Parsons, 300, 1200, 13000, 200, 1700

Wm. R. Parsons, 1200, 1200, 15400, 600, 3950

David Wheeler, 50, 80, 8000, 10, 300

Jesse Parsons, 60, 20, 1600, 15, 250

Andrew B. Parsons, 250, 50, 5000, 300, 200

Ward Parsons, 100, 500, 3000, 100, 300

Jacob H. Long, 60, 45, 1170, 75, 450

Wash. G. Long, 60, 45, 1170, 75, 450

W. Coruck, 80, 240, 2000, 80, 400

Wash. Parsons, 100, 400, 5000, 13, 400

Adam H. Long, 40, 100, 1500, 25, 450

Joab Kaylor, 140, 735, 4500, 150, 1200

Henry Jones, 30, 70, 500, 60, 300

John Bright, 12, 288, 300, 10, 150

Thos. Bright, 25, 121, 400, 10, 175

Jesse Day, 25, 225, 500, 15, 150

Ruben S. Butcher, 25, 100, 1000, 20, 225

Wm. J. Gilmore, 12, 84, 300, 5, 75

David Gilmore, 70, 35, 1200, 100, 250

Isaac A. Gilmore, 20, 80, 700, 7, 175

James R. Parsons, 490, 417, 12000, 50, 1000

Hiram Phillips, on, same, -, 5, 100

James Moore, 100,800, 2000, 25, 725

John J. Pass, 8, 160, 300, 5, 20

Jeffe M. Coruck, 25, 151, 1200, 115, 350

Abraham Parsons, 200, 1320, 800, 200, 1375

Wash. G. Coruck, 28, 107, 2430, 25, 350

Wm. D. Goff, 200, 1500, 8000, 100, 200

John R. Goff, 40, 160, 800, 30, 200

Robt. Johnson, 50, 800, 1000, 20, 200

Andrew Fansler, 80, 730, 800, 80, 500

Wm. W. Hansford, 35, 144, 700, 25, 250

Jacob Fansler, 170, 626, 3500, 150, 800

Solomon Fansler, 50, 106, 500, 20, 425

Wm. W. Parsons, 50, 147, 1000, 125, 225

Wm. E. Long, 22, -, 500, 25, 300

Garrett J. Long, 22, -, 500, 25, 300

John Flanagan, 100, 200, 3000, 50, 500

Wm. Bonner, 50, 60, 800, 5, 100

Jacob Flanagan, 75, 700, 700, 125, 450

Nathaniel Lambert, 35, 342, 4000, 15, 280

Ebanezar Flanagan, 40, 960, 2500, 20, 450

Robt. Flanagan, 30, 150, 1000, 40, 250

John Woolford, 80, 520, 1000, 20, 300

Wash. Roy, 75, 732, 2000, 20, 300

John Carr, -, -, -, 5, 175

Joab Carr, 50, 100, 1000, 5, 250

Enos Carr, 40, 110, 1000, 10, 400

Solomon Bonner, 18, 183, 400, 10, 450

James Carr, 15, 185, 300, 5, 150

Thos. L. Parsons,70, 240, 2000, 50, 350

Jacob W. Parsons, 130, 475, 3700, 30, 650

Job Parsons Sr., 150, 150, 8000, 25, 300

D. Blackman & Son, 300, 300, 8000, 50, 1768

John Dunne, 40, -, 500, 10, 150

Job Parsons Jr., 250, 150, 1000, 20,
200

Tyler County, West Virginia
1860 Agricultural Census

The University of North Carolina at Chapel Hill filmed the 1860 agricultural census for Tyler County from originals at the West Virginia State Archives under a grant from the National Science Foundation in 1963.

Columns 1, 2, 3, 4, 5, and 13 represent the following information on the census:
1. Name of Owner, Agent or Manager of Farm
2. Acres of Improved Land
3. Acres of Unimproved Land
4. Cash Value of the Farm
5. Value of Farming Implements and Machinery
13. Value of Livestock

Mark Shriver, 80, 310, 3000, 7, 226
David Chesney, 40, 64, 1500, 5, 299
George F. Fetty, 60, 100, 3000, 147, 120
Nancy Wells, 100, 150, 3000, 16, 534
Joseph Keener, 37, 3, 1000, 20, 213
Joseph Martin, 60, 30, 1000, 40, 223
Enoch Myer, 25, -, 375, 75, 41
Jonah Cooper, 30, 10, 600, 15, 30
Thomas Way, 20, 5, 375, 19, 195
Joseph Williamson, 40, 60, 1500, 50, 20
Miliam Hisam, 25, 25, 300, 15, 16
James Hisam, 9, 31, 250, 50, 360
Anderson Williamson, 50, 78, 2000, 14, 158
John W. Williamson, 80, 107, 2500, 20, 350
Esaias Pasco, 40, -, 600, 20, 129
William Weton, 100, 89, 3000, 250, 571
Peter Williamson, 30, 30, 900, 12, 271
Joshua R. Williamson, 33, 25, 1000, -, 20
Friend C. Williamson, 33, 25, 1000, -, 20

William Williamson, 190, 327, 6000, 100, 477
Benet Thorn, 25, 39, 900, 10, 285
Henry Thorn, 4, 10, 200, -, 120
James Williamson, 35, 48, 1500, 16, 222
John P. Michael, 40, 60, 1000, 15, 205
John McCandless, 50, 62, 2500, 66, 215
Rachel Martin, 80, 92, 4480, 10, 214
Abner Hisam, 36, 64, 1600, 5, 137
Rachel Hisam, 100, 118, 3000, 15, 184
Levi Hisam, 30, 70, 1000, 12, 101
Joshua R. Martin, 60, 44, 1600, 75, 242
Mordecai Morris, 100, 190, 6000, 75, 565
Isaac Lillen, 40, 146, 2000, 125, 368
John Brown, 30, -, 450, 10, 145
Joseph M. Barkheimer, 30, 72, 1000, 20, 153
Jacob Hugus, 180, 143, 4000, 165, 928
Thos. C. Williamson, 12, 50, 700, 22, 280
David Lemley, 30, 120, 1200, 12, 501

Jesse Tuttle, 45, 80, 1500, 31, 223

John G. Morgan, 180, 195, 5100, 64, 455

B. S. Morgan, 150, 165, 5000, 21, 420

Daniel Goodwin, 40, 80, 1300, 57, 387

Benjamin Wells, 340, 370, 30000, 90, 1062

Nicholas Wells, 350, 1150, 25000, 488, 1377

William Swan, 45, -, 750, 81, 355

Edward Monteith, 200, 600, 12000, 380, 898

Clawson Parker, 465, 767, 30000, 60, 1243

William Johnson, 700, 1700, 20000, 290, 3156

Peregrine Wells, 90, 224, 10000, 25, 793

John C. Parker, 35, 115, 1450, 265, 737

David S. Williamson, 70, 229, 6400, 20, 445

Thos. P. Williamson, 40, 60, 1500, 92, 218

William P. Hays, 23, 62, 1400, 10, 109

Thomas S. Steel, 40, 60, 1500, 40, 154

Adam Hays, 150, 200, 7000, 75, 244

Benj. Davenport, 200, 270, 12000, 315, 809

Joseph Shuttleworth, 38, 66, 1200, 25, 212

John W. Moore, 140, 210, 4900, 193, 938

Aaron Tichner, 42, 36, 1500, 25, 254

David Moore, 62, 37, 1200, 57, 278

Andrew Rose, 75, 75, 1800, 62, 356

Philip Y. Debolt, 30, 70, 1200, 20, 245

William C. Ash, 30, 108, 1500, 25, 242

James Smith, 65, 40, 2000, 55, 311

J. T. Hickman, 100, 290, 3000, 201, 506

John Yeck, 35, 46, 1000, 100, 272

Archibald Smith, 55, 40, 1300, 33, 277

Azel McCurdy, 25, 25, 600, 15, 121

Asa Michael, 26, 90, 800, 15, 121

Thomas Hickman, 95, 39, 1750, 53, 349

Andrew Smith, 50, 45, 1200, 52, 293

Elmore H. Fetty, 40, 76, 1500, 134, 342

William Sine, 50, 100, 2500, 81, 447

Henry Twyford, 70, 100, 2600, 69, 231

Charles Boyles, 100, 136, 4000, 80, 44

Andrew Rice, 180, 110, 3000, 400, 722

William Shook, 50, 60, 1500, 29, 186

Stephen Keener, 60, 78, 1500, 18, 288

William Rice, 210, 315, 6000, 130, 807

John Rice, 100, 200, 3000, 146, 475

Joseph Lightner, 100, 100, 2000, 61, 411

John Wade, 60, 67, 1400, 25, 159

Alexander McLaughlin, 40, 31, 1200, 88, 275

James C. Williamson, 130, 170, 7000, 56, 434

John Ankrom, 30, 63, 3000, 180, 436

Richard Ankrom, 120, 437, 12000, 250, 493

William Robinson, 35, 15, 600, 20, 65

Joshua Russell, 70, 268, 5000, 205, 1000

Barney Wells, 300, 1000, 38800, 400, 2700

John Tuttle, 100, 157, 2000, 60, 404

Joseph Sandy, 40, -, 600, 85, 427

Henderson Sandy, 30, -, 450, 75, 260

Thomas H. Stewart, 30, 45 1500, 125, 233

John B. McCoy, 75, 75, 7500, 300, 828

Stephen Watkins, 150, 200, 3000, 150, 659

Thomas Kyser, 50, 50, 2000, 83, 347

Jacob Maginnis, 40, 60, 600, 35, 149

Benjamin Shriver, 150, 188, 8000, 150, 609

George Lewis, 70, 30, 2500, 205, 495

James E. Engle, 30, 85, 2000, 60, 224

George Scott, 130, 99, 4000, 150,506

Nevin Porter, 140, 127, 5340, 100, 781

Joseph Wade, 110, 126, 3500, 55, 478

Noble Stewart, 60, 43, 1800, 100, 350

Helim Fish, 50, 50, 2000, 100, 213

Samuel Cox, 40, 35, 800, 25, 175

Robt. Cunningham, 50, 50, 1000, 35, 418

David Roberts, 75, 203, 3200, 130, 227

James Coffield, 70, 80, 1500, 75, 342

Daniel Lee, 14, 11, 150, 25, 110

William Roome, 80, 30, 2800, 95, 507

Henry W. Shook, 80, 30, 2000, 125, 465

Iret Jenkins, 20, 70, 700, 15, 45

Joseph Jenkins, 35, 85, 1800, 50, 305

Josephus Cooper, 20, 10, 600, 25, 380

Solomon Slider, 16, 16, 300, 20, 134

John Billiter, 40, 60, 1000, 60, 170

William Kemble, 100, 100, 2000, 25, 570

James Kemble, 20, 94, 1000, 3, 257

William Buck, 25, 28, 900, 50, 238

Thomas Richerson, 100, 104, 3000, 60, 317

Samuel McCoach, 65, 40, 1200, 55, 382

James P. Michael, 20, 30, 700, 25, 212

John McCoach, 75, 58, 1600, 70, 414

Thomas Coffield, 40, 22, 625, 75, 197

John Christen, 65, 35, 1000, 60, 300

William McCormick, 65, 41, 2000, 84, 508

Isaac Rice, 100, 62, 2430, 90, 410

Abram Kimble, 75, 25, 1500, 30, 267

James Covalt, 50, 58, 1200, 20, 147

James Peoples, 100, 100, 3500, 100, 791

William Corbitt, 120, 100, 4000, 150, 678

Robert Corbitt, 120, 100, 4000, 150, 678

John Scarlott, 55, 159, 3000, 50, 262

Omer Granden, 40, 60, 1200, 25, 113

Daniel Sweeney, 75, 37, 1344, 30, 748

John Jobes, 45, 60, 1500, 20, 249

Uriah Ice, 50, 60, 1500, 24, 242

Gains McRoberts, 30, 40, 800, 76, 82

Thomas Keller, 60, 258, 3180, 80, 492

Martin Keller, 50, 113, 1630, 56, 357

James Evans, 32, 120, 1500, 20, 190

Milton Tallman, 40, 308, 2500, 60, 160

Isaac Holmes, 60, 200, 5000, 100, 421

James C. McCoy, 75, 188, 7000, 125, 650

Aulds Bullah, 50, 125, 2000, 64, 280

Hiram Venham, 40, 40, 650, 30, 323

Thomas S. Lacy, 100, 350, 4000, 200, 653

Jacob Williams, 60 70, 1400, 40, 243

Peter Thomas, 50, 50, 700, 25, 274

James E. Stewart, 50, 50, 100, 50, 215

Jacob McCoy, 75, 104, 3000, 50, 423

William Archer, 200, 182, 6000, 75, 633

Joseph Archer, 40, 53, 700, 50, 463

Neil Archer, 150, 147, 6000, 109, 718

Arthur Archer, 114, 176, 4400, 100, 432

Joseph McCoy, 90, 101, 4500, 50, 536

Joseph Gorrel, 75, 75, 1000, 85, 725

John B. Gorrel, 325, 319, 1200, 100, 807

Zackquill Pierpoint, 100, 189, 2890, 60, 619

William Ice, 50, 90, 1800, 30, 275

William Buck, 70, 178, 2480, 50, 661

Alexander Buck, 40, 60, 2000, 30, 299

John Buck, 40, 135, 900, 50, 297

Henry Buck 20 80, 800, 16, 129

George T. Haslep, 100, 93, 2500, 25, 331

Isaac Ice, 70, 140, 800, 50, 208

Laban Mercer, 150, 100, 2500, 125, 558

Thomas Bucher, 50, 950, 3000, 30, 244

Jonathan Jacob, 50, 89, 1500, 55, 290

James Ferrel, 100, 300, 3500, 200, 578

Eli Ferrel, 60, 140, 2000, 50, 276

Abner Pipes, 50, 100, 1500, 30, 277

Jacob Dervan, 30, 10, 500, 20, 305

Ralph Prickett, 25, 90, 1800, 60, 265

Elizabeth Prickett, 60, 200, 2000, 50, 160

Nathaniel T. Riggs, 120, 80, 4000, 45, 580

Simeon H. Hocking, 150, 500, 4750, 250, 676

John C. Beaty, 226, 1391, 11330, 167, 827

William Prichard, 80, 90, 3000, 49, 700

James Jenkins, 50, 110, 200, 75, 260

Arthur Ankrom, 150, 250, 6000, 75, 1023

James Hill, 75, 225, 2500 50, 469

Mansfield Stealey, 160, 360, 8325, 65, 74

Joseph Thomas, 70, 142, 4000, 125, 206

Joseph Lazear, 150, 130, 3500, 80, 780

Jesse C. Lazear, 125, 10, 4000, 70, 597

William Thomas, 100, 62, 2500, 220, 302

Edmund Brannon, 70, 40, 2000, 20, 298

David Miller, 50, 20, 1000, 75, 305

John Lady, 25, 16, 800 20, 226

John Keller, 150, 180, 3500, 50, 628

Jesse A. Dancer, 200, 120, 6975, 110, 1408

John Thornburn, 90, 99, 3000, 25, 314

David Hisam, 100, 132, 2000, 30, 341

Ephraim Martin, 60, 75, 1500, 25, 213

Granville Flesher, 75, 75, 2500, 38, 226

Jackson Morgan, 95, 38, 1250, 31, 254

John Craig, 170, 150, 6000, 80, 1088

Peter Woodburn, 250, 325, 12000, 125, 678

Meshach Baker, 80, 41, 1800, 75, 288

Thomas Pierpoint, 75, 74, 4000, 75, 457

Francis Pierpoint, 40, 60, 1500, 100, 230

Josephus Roberts, 75, 25, 3000, 105, 778

Peter Hartley, 125, 65, 4000, 150, 818

Jacob Stealey, 70, 80, 1800, 70, 132

Jeremiah Keck, 30, 70, 2000, 25, 166

William Pierpoint, 75, 141, 1200, 20, 223

Joseph Reed, 75, 88, 2500, 52, 437

John Kerr, 175, 475, 7000, 100, 210

J. T. Nicklin, 19, 14, 2400, 70, 315

Henry A. Rymer, 120, 3654, 8300, 49, 394

William W. Boreman, 50, 5000, 21000, 133, 457

David Hickman, 45, 5939, 14500, 100, 330

Christian Engle, 25, 800, 3000, 50, 365

James Stealey, 490, 1000, 2600, 250, 2150

Anthony Asher, 75, 150, 2000, 46, 543

John B. Smith, 80, 390, 5000, 25, 627

William W. Gorrel, 70, 210, 3000, 95, 532

Daniel Anderson, 26, 120, 1000, 45, 286

T. F. Lazear, 200, 85, 4000, 150, 1100

John McKay, 150, 425, 7500, 80, 800

William Garman, 70, 97, 2500, 50, 466

Henry Masters, 30, 26, 500, 25, 176

William Stewart, 60, 188, 1800, 30, 547

Joseph M. Holmes, 120, 83, 5500, 200, 636

Thomas Swan, 30, 9, 800, 20, 166

Lewis M. Marsh, 275, 300, 10000, 200, 1311

John Lazear, 100, 100, 3000, 16, 67

Lewis W. T. Martin, 40, 10, 1000, 28, 308

Franklin Warman, 40, 60, 950, 150, 300

Jeremiah M. Martin, 20, 83, 625, 25, 128

Henry Metz, 35, 65, 1200, 43, 301

Reuben Martin, 125, 110, 3000, 125, 661

Zeri Workman, 40, 66, 1200, 25, 247

William Travis, 80, 60, 2000, 90, 745

Stephen Buchanan, 70, 30, 1500, 100, 437

Manning Martin, 120, 40, 2000, 100, 681

William Camp, 70, 41, 1500, 38, 395

Aaron Cornell, 70, 89, 2000, 50, 375

Benjamin Cornell, 90, 54, 2500, 35, 295

Thomas Smith, 30, 50, 4000, 50, 150

Eliza Smith, 25, 105, 2000, 35, 310

John Smith, 15, 35, 600, 25, 201

Joseph M. Morgan, 162, 219, 7000, 81, 676

Jacob Patterson, 50, 100, 1200, 30, 281

Elias Wells, 40, 111, 1500, 25, 140

Elizabeth Gorrel, 100, 150, 3000, 34, 291

George Gorrel, 70, 89, 2000, 109, 436

Elias Gorrel, 242, 1018, 18625, 125, 1209

John Smith, 200, 172, 5000, 80, 630

William K. Smith, 40, 127, 1600, 25, 262

Jacob T. Galaway, 240, 160, 6500, 200, 546

F. H. Flesher, 90, 185, 4000, 80, 443

Eli Flesher, 65, 185, 3000, 100, 447

John Bullman, 85, 128, 3000, 85, 346

Andrew Bullman, 55, 125, 1300, 25, 282

Jacob Ankrom, 50, 90, 1200, 30, 218

Elizabeth Mercer, 50, 170, 1800, 25, 293

Thomas Steel, 70, 330, 4000, 50, 619

Robert Allison, 30, 45, 1000, 30, 157

Henry Copehaver, 40, 116, 2000, 100, 228

Abraham Lamp, 30, 120, 2000, 20, 192

John Hamlin, 100, 300, 3000, 105, 376

Charles Williams, 40, 21, 600, 25, 235

James Barker, 150, 150, 2000, 100, 682

Jacob Bradford, 75, 150, 1500, 30, 278

A. J. Joseph, 70, 10, 1448, 70, 351

John Hammond, 200, 440, 6000, 250, 1092

Horace Hill, 150, 377, 3000, 140, 506

Benjamin Hill, 50, 162, 1200, 30, 314

Henry Bradford, 75, 107, 900, 100, 241

Samuel Howard, 120, 142, 3700, 108, 940

Joseph Owen, 45, 158, 1400, 20, 268

John Carder, 30, 112, 1500, 30, 233

Nelson Ankrom, 60, 290, 2000, 20, 407

Eli Davis, 40, 60, 1000, 25, 215

Eli B. Long, 30, 45, 600, 25, 195

George Long, 200, 555, 6000, 150, 664

Charles Eberhart, 100, 283, 6000, 135, 636

Anthony Smith, 100, 200, 3000, 35, 460

Jeremiah Bedlman, 100, 64, 2500, 40, 567

John S. Ewell, 60, 59, 1250, 15, 195

Enos Smith, 100, 430, 4000, 50, 521

Levi Starkey, 30, 40, 500, 15, 257

Charles Howard, 25, 175, 800, 25, 182

Samuel Moore, 25, 73, 900, 16, 147

Wm. Robinson, 60, 55, 1200, 120, 363

Alvan Robinson, 70, 130, 1600, 55, 215

Abraham Shriver, 70, 195, 2650, 106, 293

Abram Umersetter, 34, 26, 550, 16, 190

Rolly Moore, 50, 31, 1200, 55, 332

Wilson Long, 80, 175, 2530, 70, 317

Joseph Philips, 28, 22, 600, 20, 172

Richard Stealey, 40, 60, 1200, 25, 270

Titus Philips, 30, 10, 400, 16, 105

Aaron Darnel, 40, 35, 800, 12, 138

John S. Hare, 40, 60, 1000, 25, 289

Michael Frank, 70, 107, 2500, 20, 348

Kenner Smith, 150, 175, 3075, 40, 284

Isaac R. Smith, 48, 50, 1200, 20, 214

Ralph Smith, 75, 105, 2500, 75, 490

O. P. Stealey, 12, 95, 1000, 25, 108

Anthony Ash, 25, 75, 500, 20, 237

John J. Ash, 70, 184, 2000, 30, 657

John Ash, 90, 400, 5000, 25, 498

Peter Smith, 30, 260, 1500, 40, 293

Benj. Freeland, 100, 350, 2000, 45, 428

Peter D. Ash, 50, 200, 1600, 85, 307

Isaac Davis, 65, 335, 2800, 100, 355

Barney W. Fox, 30, 220, 1600, 40, 300

Joshua Craven, 70, 190, 1500, 50, 316

Levi Grim, 80, 73, 1500, 50, 346

James Wright, 50, 32, 1000, 25, 237

Saml. Vandergrift, 25, 75, 700, 75, 188

Andrew W. Duty, 75, 500, 4000, 50, 422

Lemuel Riggs, 70, 90, 1000, 100, 368

Wm. McCullough, 40, 57, 1000, 45, 308

Elmer Riggs, 35, 100, 1200, 20, 201

Thomas Grim, 65, 135, 700, 34, 430

John Thomas, 100, 90, 3000, 90, 344

Margaret Bond, 80, 121, 2500, 105, 257

Ashberry Varner, 100, 25, 2000, 100, 492

John S. Smith, 35, 75, 1500, 30, 250

Amos Headley, 80, 63, 1700, 100, 334

Isaac Smith, 220, 480, 11500, 250, 1411

John W. Hamilton, 25, 51, 1000, 16, 124

Jeremiah Hoskinson, 60, 100, 1900, 30, 312

Andrew S. Corbly, 85, 255, 2500, 120, 375

Barney Bond, 50, 105, 1540, 30, 369

Daniel McCullough, 35, 65, 800, 25, 295

William Smith, 30, 70, 900, 30, 150

Thomas Bond, 140, 260, 5000, 140, 532

Butler B. Bond, 40, 90, 1200, 25, 208

Hamilton Doak, 30, 80, 900, 16, 153

Hiram M. Doak, 40, 200, 2500, 15, 204

Charles H. Conaway, 100, 80, 4000, 150, 334

William Wells, 150, 75, 6000, 100, 869

Alpheus Conaway, 250, 500, 7000, 300,845

William N. Moore, 60, 37, 1200, 40, 301

William Moore, 65, 15, 2000, 60, 188

Charles Davis, 70, 180, 3000, 80, 457

Henry Lemly, 50,185, 1200, 45, 303

William Smith, 10, 30, 500, 25, 135

Caleb Perkins, 100, 130, 3000, 300, 421

Waitman Joseph, 290, 1948, 17040, 300, 1970

Alexander Ullum, 50, 85, 1500, 60, 2150

Z. B. Riggle, 200, 353, 16000, 100, 745

Robert W. Davis, 50, 200, 3000, 30, 369

Alexander M. Davis, 60, 36, 1600, 200, 496

Nathan Joseph, 100, 260, 2500, 200, 542

Alexander Doak, 90, 175, 3000, 40, 417

William Waters, 40, 30, 800, 30, 287

Calvin Hanes, 35, 500, 2000, 25, 127

Lorenzo Cane, 30, 350, 1500, 100, 294

Jacob Weekley, 100, 108, 6000, 86, 432

John T. Dawson, 70, 46, 1500, 55, 265

Robert Davis, 40, 145, 1000, 30, 222

Nathaniel Smith, 50, 80, 1400, 18, 169

Robert Scott, 50, 70, 1600, 30, 250

John Underwood, 40, 70, 1500, 40, 160

William James, 50, 88, 1650, 40, 210

Nicholas Orr, 50, 50, 2000, 80, 350

Daniel Weekley, 85, 48, 1500, 30, 402

Robert Davis, 70, 55, 12000, 60, 192

E. M. B. Keys, 50, 70, 1000, 20, 337

Edward A. Montgomery, 40, 65, 1500, 50, 69

Robert Doak, 100, 200, 3000, 50, 426

Cephas Duckworth, 125, 125, 3500, 55, 480

John Gregg, 70, 15, 2000, 75, 309

Jacob Stewart, 70, 30, 1400, 80, 345

Samuel Doak, 60, 100, 1500, 30, 334

George Bond, 183, 131, 8000, 40, 612

John Ireland, 150, 307, 6000, 250, 637

Lewis Ripley, 30, 218, 1000, 16, 155

William Jemason, 180, 604, 9012, 275, 965

John Smith, 50, 150, 2000, 25, 343
Alfred Smith, 50, 150, 2500, 40, 360
Henry Smith, 50, 150, 2500, 85, 393
James Smith, 100, 210, 4000, 40, 453
Thomas Weekley, 30, 20, 500, 16, 154
Joseph Twyman, 20, 80, 750, 16, 59
Hugh Smith, 75, 447, 4100, 65, 517
Isaiah Baker, 80, 250, 3000, 60, 497
Robert Anderson, 50, 64, 2000, 35, 380
Isaac Noles, 15, 35, 300, 15, 211
Thomas Conaway, 80, 940, 3500, 150, 509
John B. Ferrel, 35, 222, 3000, 25, 232
Silas Tenant, 30, 150, 1500, 20, 265
Latham Inghram, 40, 164, 1500, 20, 232
Benjamin Haught, 40, 175, 2000, 16, 305
Andrew Haught, 40, 80, 1500, 23, 347
Joshua Haught, 15, 85, 500, 16, 201
Eli Parks, 30, 120, 1000, 35, 219
James Elder, 30, 70, 900, 16, 148
Eugenius Lamasters, 40, 112, 1800, 75, 500
Alexander Tenant, 50, 198, 1736, 60, 366
Jacob H. Lyons, 30, 187, 1000, 56, 147
Enoch Lamasters, 75, 217, 1700, 35, 345
Enoch Bailey, 60, 80, 1000, 80, 349
Elias Nichols, 100, 330, 3000, 100, 690
Elijah Nichols, 30, 120, 1000, 20, 253
Elijah Waters, 40, 47, 900, 20, 177
Hamilton G. Frum, 80, 134, 1700, 30, 225
William Baker, 40, 207, 1300, 10, 134
John Elder, 90, 195, 2000, 80, 508

Henderson McIntire, 40, 160, 1000, 15, 160
William Way, 20, 30, 460, 20, 360
Ezekiel Pitts, 30, 97, 1000, 50, 320
Alfred Spencer, 180, 920, 5500, 100, 430
Levi Weekley, 50, 96, 1500, 47, 306
Sampson Booher, 35, 115, 1000, 20, 180
Thomas Woodbourn, 70, 170, 2000, 25, 390
Catharine Booher, 80, 500, 2000, 25, 470
Thomas S. Kester, 30, 60, 700, 12, 168
Postly Booher, 30, 40, 1000, 16, 130
William Baker, 40, 90, 1240, 60, 303
John Weekly, 20, 80, 1000, 20, 222
James Weekly, 60, 140, 2000, 20, 229
Samuel Woodbourn, 70, 330, 1800, 30, 205
Jacob Lantz, 85, 575, 3600, 30, 334
Henry Stoneking, 50, 65, 1000, 70, 270
Joseph Tenant, 100, 220, 3000, 90, 401
Andrew Thompson, 50, 370, 4000, 90, 402
Rawley Lamasters, 60, 140, 1200, 50, 229
Andrew Tusten, 70, 134, 1216, 20, 226
Jacob Tusten, 80, 254, 3000, 30, 341
Elisha Hensley, 70, 100, 1200, 25, 459
John S. Glover, 50, 50, 1000, 25, 181
John W. Stackpole, 60, 135, 1800, 100, 208
Dalia Lamasters, 55, 151, 1958, 40, 444
Septimus Lamasters, 100, 393, 4000, 100, 792
William Main, 50, 157, 1000, 50, 427

William Lyons, 70, 730, 6200, 100, 204

Richard Weekley, 100, 90, 4000, 100, 698

Joab McIntire, 40, 240, 1200, 30, 259

Peter Moore, 300, 822, 9000, 200, 732

George McIntire, 30, 217, 1050, 30, 269

James McIntire, 40, 184, 1200, 30, 260

Jacob McIntire, 40, 182, 1500, 40, 376

Jonathan Wright, 60, 850, 700, 125, 578

Joseph McCormick, 170, 930, 7200, 75, 492

John White, 80, 379, 6000, 75, 381

Isaac Underwood, 30, 170, 1000, 60, 346

Thomas Smith, 16, 84, 800, 30, 266

Absalom George, 200, 300, 4000, 100, 937

Samuel Underwood, 200, 550, 9600, 250, 900

John Underwood, 115, 350, 3000, 16, 393

Mary Sandy, 50, 200, 2000, 20, 344

Robert D. Ferrel, 50, 25, 1000, 25, 291

Solomon Underwood, 40, 80, 1000, 30, 450

Jonathan Underwood, 35, 65, 1000, 30, 371

Isaac Underwood, 75, 177, 4000, 100, 496

William Pratt, 75, 275, 2000, 75, 392

Wm. Underwood, 140, 250, 5500, 100, 594

Robert Orr, 80, 190, 3000, 60, 893

William W. Clark, 80, 527, 2250, 100, 418

Felix Watts, 19, 11, 300, 30, 296

Hiram Sweeney, 80, 190, 8000, 25, 202

Bowers Furbee, 80, 150, 5000, 100, 499

John Hustead, 60, 128, 2000, 70, 234

Philip Seckman, 130, 170, 4600, 89, 454

Andrew Seckman, 75, 118, 4000, 110, 587

Andrew J. Furbee, 80, 195, 4000, 70, 400

James M. Smith, 250, 450, 10000, 200, 1138

Perry Pratt, 25, 71, 500, 20, 160

James Underwood, 50, 120, 1500, 50, 425

Samuel Underwood, 50, 90, 2100, 75, 246

Oliver Wells, 350, 900, 11650, 150, 4193

Daniel Sweeney, 100, 48, 3500, 30, 246

Jacob Thomas, 80, 104, 4000, 60, 378

John Graves, 80, 220, 3000, 100, 250

Henry Thomas, 25, 75, 600, 25, 385

Jonathan Ankrom, 100, 54, 3000, 100, 100

William Gregg, 150, 650, 12800, 80, 698

Robert Stather, 100, 50, 2600, 75, 474

Thomas Smith, 100, 100, 4000, 150, 565

Albert Fletcher, 45, 27, 1600, 20, 105

George Mason, 320, 593, 9564, 250, 960

Jesse Henderson, 100, 84, 2700, 150, 501

Andrew D. Conaway, 125, 225, 4500, 150, 934

William Stealey, 250, 216, 8800, 75, 1456

Gilbert Smith, 100, 100, 3000, 100, 541

Armstrong Smith, 70, 255, 3000, 75, 483

John Crumrine, 100, 72, 4000, 150, 437

Kenner A. Smith, 120, 280, 5000, 150, 428

James Morris, 150, 394, 8000, 200, 661

Samuel Nicklin, 100, 120, 3000, 150, 470

William Ankrom, 75, 25, 2500, 75, 391

Mary Ripley, 75, 25, 2000, 200, 340

Daniel Ripley, 70, 84, 2500, 85, 541

John Fordyce, 100, 142, 2000, 20, 401

Stephen V. Wilcox, 70, 299, 1845, 20, 233

William Nicklin, 28, 2, 550, 20, 80

James Taggart, 100, 74, 2000, 180, 520

Thomas Wheeler, 45, 111, 1400, 25, 267

Upshur County, West Virginia
1860 Agricultural Census

The University of North Carolina at Chapel Hill filmed the 1860 agricultural census for Upshur County from originals at the West Virginia State Archives under a grant from the National Science Foundation in 1963.

Columns 1, 2, 3, 4, 5, and 13 represent the following information on the census:
1. Name of Owner, Agent or Manager of Farm
2. Acres of Improved Land
3. Acres of Unimproved Land
4. Cash Value of the Farm
5. Value of Farming Implements and Machinery
13. Value of Livestock

J. W. Shepherd, 40, -, 210, 20, 100
Levi J. Queen, 10, 90, 300, 10, 100
J. H. Jimieson, 60, 40, 1500, 20, 400
B. W. Waugh, 10, -, 200, 25, 230
C. J. Rigney, 10, -, 200, 10, 150
P. L. Tomblin, 6, -, 120, 5, 50
Jacob Owen, 80, 190, 3500, 75, 500
Isaac Brake, -, -, -, -, -
Melville Brake, 60, 40, 2500, 100, 900
Jno. N. Loudin, 40, 34, 1200, 15, 250
Jas. Dix, 180, 230, 4500, 40, 800
Jno. Mick, 75, 100, 2500, 30, 550
Jas. L. Jennings, 10, -, 200, 12, 135
Elias Cayner, 27, -, 400, 10, 150
Howard Roan, 7, 33, 100, 1, 25
Augt. W. Sexton, 60, 110, 2000, 10, 200
S. D. Barr, 15, -, 300, -, 30
Sol. Suter, 10, -, 200, 2, 50
Joshua Shultz, 5, -, 100, -, 60
John Ireland, 100, 100, 5000, 100, 800
Cyrus Chineworth, 10, -, 150, -, 25
Adam Gowen, 10, -, 200, 5, 25
A. R. Ireland, 300, 400, 16000, 75, 1250
Julius Vaulters, 8, -, 160, -, 30
Andrew Robinson, 10, -, 250, -, 20
Wm. Loudin, 75, 45, 2000, 5, 200

Thos. Loudin, 100, 100, 4000, 75, 525
Jno. M. Loudin, 50, 50, 2000, 20, 400
Wm. Paugh, 40, 56, 1600, 20, 300
Wm. S. Loudin, 30, 310, 2500, 5, 125
G. F. Browning, 10, 40, 600, 5, 100
Mary Mick, 25, 8, 300, 5, 125
Nicholas Mick, 70, 300, 2500, 20, 450
A. G. Reeder, 35, 105, 1800, 5, 250
H. H. Mayes, 15, 10, 350, 3, 30
Willis Mayes, 6, -, 120, -, 20
E. A. Mick, 30, 41, 800, 5, 100
Isaac Owens, 50, 131, 2200, 150, 350
Levi Hotsepiller, 55, 180, 2500, 75, 250
J. L. D. Brake, 30, 60, 1000, 20, 300
J. J. Campbell, 10, -, 200, -, 20
Jac. Rodbaugh, 25, 36, 550, 8, 125
Thos. Bice, 20, 20, 600, 2, 50
W. B. Loudin, 30, -, 500, 5, 230
F. Reeder, 13, 104, 600, 15, 200
Chas. Mick, 130, 130, 3000, 60, 450
Isaac Rodbaugh, 30, 68, 1500, 5, 120
C. Herenden, 14, 85, 600, 5, 20
Mary Lewis, 40, -, 1000, -, 200
John Lewis, 62, 62, 1000, 75, 400

Sol. George, 10, 30, 200, 2, 6
Mic. Hoover, 18, 32, 300, 4, 85
J. L. Keesling (Kusling), 55, 25, 1400, 10, 225
B. B. Musgrave, 90, 110, 3200, 25, 500
E. Cook, 40, 57, 1800, 5, 100
J. Mowery Jr., 60, 15, 1500, 25, 350
Joul Hartman, 50, 50, 2000, 25, 200
Ant. & J. H. Rohrbaugh, 120, 30, 4300, 100, 600
Eph. Adams, 75, 30, 1800, 80, 280
Reuben Lough, 6, -, 120, 12, 35
John Hinkle, 150, 50, 3000, 50, 500
M. Ritter, 30, 60, 1000, 50, 200
Benj. Miller, 30, -, 250, -, 35
Eliz. Jackson, 150, 300, 4000, 25, 250
A. Carper, 120, 50, 4500, 100, 800
Benj. Carper, 200, 100, 7500, 20, 800
Isaac Post, 450, 250, 14000, 50, 1350
Job Hinkle, 335, 195, 10200, 100, 1350
A. M. Bastable, 320, 9000, 21500, 175, 1200
J. W. Browning, 10, 120, -, 5, 25
E. Fornash, 3, -, 75, -, 30
Benj. Rodebaugh, 320, 90, 8800, 75, 1320
Job Ward, 40, 60, 1000, 20, 120
George Post, 365, 335, 17000, 100, 1450
David A. Casto, 80, 37, 2000, 25, 500
Jno. H. Crites, 18, -, 300, 15, 110
Jacob Teter, 150, 50, 5000, 100, 950
Silas Bennett, 115, 65, 3500, 100, 400
Walter Loudin, 95, 100, 2500, 50, 550
O. B. Loudin, 75, 45, 2000, 10, 340
Abr. Bennett, 10, 400, 900, 12, 150
Amz. Dawson, 40, 105, 2000, 20, 170

Wm. Frymier, 20, -, 300, 12, 150
Joseph Ward, 30, 117, 1600, 20, 200
W. W. Warner, 20, 97, 1200, 18, 120
Simon Roberts, 28, -, 300, 2, 350
D. C. Loudin, 8, -, 150, 22, 60
Wm. A. Smith, 3, -, 30, 12, 30
S. W. Bennett, 80, 10, 1600, 20, 250
Moses Reeder, 35, 55, 800, 20, 225
D. M. Bennett, 8, -, 160, 12, 25
J. M. E. Berria, 6, -, 120, 14, 150
Sus__ Depoy, 75, 170, 2500, 10, 200
M. H. Bennett, 4, -, 80, 12, 25
Elias Bennett, 120, 30, 3000, 100, 690
Jos. Register, 55, 55, 1600, 20, 250
Isaac Warner, 75, 140, 1800, 15, 100
Owen Westfall, 10, -, 150, -, 40
Jas. Gowers, 4, -, 60, 20, 100
A. Register, 5, -, 70, -, 35
J. J. Post, 125, 175, 4000, 100, 500
Stephen Post, 100, 146, 3000, 25, 500
Abr. J. Post, 300, 500, 9800, 100, 2000
Geo. Hoover, 18, 3, 250, 15, 200
J. B. Queen, 10, 115, 750, 50, 150
Jac Oldakers, 30, 31, 600, 20, 100
Ant. Strader, 180, 190, 5800, 100, 300
Joel Casto, 80, 285, 3500, 45, 380
Tho. Kyle, 5, -, 100, 2, 25
Jacob Post, 30, 130, 1600, 20, 150
John Lance, 12, -, 180, 12, 100
Adam White, 10, -, 150, 22, 150
Nic. McVaney, 60, 80, 1600, 30, 250
Saml. Neeley, 8, -, 160, 14, 60
Jac. Starcher, 100, 94, 3000, 50, 885
Isaac Hinsman, 200, 150, 9000, 50, 2600
Mic. Murphrey, 8, -, 120, 12, 80
John A. Peters, 35, 80, 1600, 20, 225
Jos. Lowther, 4, -, 80, -, 20
Jas. M. Gulley, 10, -, 150, -, 50
Benj. Stout, 75, 100, 2000, 170, 600
C. G. Norman, 30, 46, 1400, 20, 150
D. T. Talbott, 60, 105, 1600, 15, 300

Richd. Thrash, 60, 58, 1200, 5, 150
Elias M. Queen, 10, -, 150, 18, 150
John Bear, 10, 57, 700, 15, 160
Josh. Lewis, 75, 70, 2600, 20, 150
L. H. Queen, 70, 40, 2000, 20, 300
Geo. W. Queen, 10, -, 200, 10, 75
P. G. Smith, 10, 16, 350, 12, 50
Richd. Oldakers, 18, 48, 700, 15, 200
Jasper McPherson, 60, 31, 1100, 15, 350
Abr. Oldakers, 80, 70, 1600, 20, 375
Wm. Oldakers, 60, 78, 1600, 20, 300
Saml. Davis, 60, 142, 2800, 100, 400
Jno. A. Sheets, 75, 86, 2100, 40, 630
Granville Queen, 6, -, 120, 10, 150
Jos. Flint, 150, 50, 3500, 75, 420
Danl. Flint, 5, -, 100, 10, 25
Henry Baughman, 25, 33, 1000, 20, 150
Wm. Beasly, 20, 27, 700, 5, 120
Brumfield Bond, 120, 280, 6000, 15, 800
Boothe Bond, 7, 20, 300, 14, 320
Hez. Hess, 5,-, 100, -, 18
Amos Reeder, 40, 10, 800, 5, 175
D. D. Curtis, 25, 25, 900, 15, 125
H. L. Curtis, 35, 23, 900, 15, 175
Stephen Curtis, 215, 35, 800, 15, 120
J. M. Jemison, 5, -, 100, 10, 50
Jno. W. Curtis, 15, 85, 600, 12, 100
Jno. W. Marple, 220, 340, 11500, 120, 1000
Alb. Marple, 15, -, 300, 15, 300
Ad. Marple, 10, -, 200, 15, 200
F. Heavener, 5, -, 100, 5, 25
D. B. Reger, 175, 187, 6000, 100, 1100
John Paugh, 8, -, 150, 5, 125
John Warner, 15, -, 200, 10, 75
Lewis Karrikoff, 145, 132, 9250, 150, 1050
Thos. Swick, 10, -, 100, 5, 50
Jesse Cummings, 10, 200, -, 25, -
Joseph Carper, 270, 30, 4000, 25, 1170

Geo. Westfall, 120, 116, 6500, 20, 350
Enoch Westfall, 10, -, 200, 20, 150
Benj. Conley, 53, 503, 1760, 95, 350
Jackson Warner, 40, 60, 1500, 25, 120
John Ward, 10, -, 200, 5, 175
Jno. W. Casto, 75, 42, 2700, 40, 335
D. J. Brake, 30, -, 50, 15, 150
J. Q. Harvey, 25, -, 300, 15, 250
Job Casto, 50, 63, 2000, 75, 450
S. G. Martin, 26, 25, 800, 15, 150
J. M. Matheny, 30, 17, 900, 15, 400
W. S. Gum. 75, 61, 2160, 80, 300
WM. Shultz, 20, 50, 1200, 13, 150
Benj. Harvey, 10, -, 200, 13, 100
Joel Life, 5, -, 100, 12, 25
Diad. Stump, 75, 76, 1900, 25, 200
Jac. Carpenter, 10, -, 200, -, -
Nelson Robinson, 25, 327, 9300, 100, 930
J. L. Queen, 10, -, 200, 15, 150
Christo. Crites, 10, -, 200, 13, 100
Jonas Crites, 10, -, 150, 10, 35
David Wilson, 20, -, 300, 22, 75
Geo. Simons of J., 50, 334, 3500, 75, 220
W. H. Gregory, 10, -, 200, 10, 25
J. C. McCloud, 10, -, 200, 10, 125
David Reed, 8, -, 160, 5, 250
Jacob Reed, 5, 85, 1200, 15, 200
John Matheny, 45, 23, 1000, 20, 220
Jos. Matheny, 33, 20, 1000, 40, 170
Jas. J. White, 12, 10, 200, 12, 40
Stewart L. Queen, 200, 160, 7000, 200, 1200
M. Hannah, 20, 56, 900, 15, 40
B. R. Simons, 10, -, 200, 12, 100
Pat. Moran, 50, 54, 1200, 15, 180
A. Lanham, 6, -, 75, 10, 40
Wm. Jenkins, 50, 162, 3500, 50, 485
M. Robinson, -, 145, 225, 4, 10
A. Grimes, 10, -, 200, -, 10
Andrew Gum, 55, 130, 1800, 25, 400
Johnson Arnold, 110, 250, 6000, 45, 565

A. J. Courtney, 10, -, 200, -, 40
Al. Matheny, 15, -, 300, 13, 150
Jno. Fisher, 10, -, 200, 13, 125
John Love, 80, 237, 5000, 60, 650
Alexr., Reed, 10, -, 200, 13, 110
Roswell White, 10, 18, 200, 10, 25
M. Warner, 5, -, 100, -, 25
Ab__ Queen, 50, 25, 500, 15, 395
J. W. Alexander, 5, -, 100, -, 100
S. W. Metheny, 8, -, 100, 12, 30
Wm. Floyd, 40, 60, 2000, 70, 400
Jacob Wagner, 12, 16, 320, 12, 60
Wat. Reynolds, 30, 30, 750, 25, 325
J. W. Metheny, 7, -, 100, 13, 150
Jos. Reynolds, 16, 12, 350, 13, 70
Jonas Cooper, 45, 15, 900, 25, 300
Laft. Hinkle, 80, 47, 1800, 45, 475
Geo. Hoover, 10, -, 200, 10, 125
Phil. Depoy, 120, 94, 4400, 90, 565
M. Boatright, 60, 100, 3000, 20, 300
Asa Crites, 8, -, 100, 10, 70
Joab Crites, 20, 20, 600, 13, 130
Jos. Lance, 10, -, 200, 15, 200
Jacob Brake, 100, 100, 3500, 30, 200
Jno. Kuykendall, 10, 200, 200, 10, 50
Edmd. Fitzgerald, 15, -, 200, 5, 175
Wyatt Fitzgerald, 10, -, 200, 20, 100
Geo. Lance, 8, 4, 500, 12, 200
M. Lance, 8, -, 100, 1, 120
J. C. Alderman, 8, -, 120, 10, 110
Alva Setn, 280, 244, 10000, 200, 1850
John Reger, 120, 1160, 5200, 100, 400
D. Reger, 8, 100, 800, 25, 375
Mic. Strader, 200, 262, 7500, 150,800
Chrs. Post, 20, -, 300, 15, 200
W. W. Foster, 30, 72, 500, 25, 120
G__ Strader, 86, 50, 1800, 75, 500
Mic. Strader Jr., 52, 80, 1800, 15, 450
Morgan Wilfong, 10, -, 150, 12, 38
N. J. Fronsman, 30, 35, 650, 25, 130
Jos. Lewis, 20, -, 200, 125, 130

S. Rohrbaugh, 200, 30, 5000, 160, 425
Danl. Post, 280, 240, 9500, 200, 1400
Nicholas Post, 10, -, 200, 25, 200
Thos. Dean, 10, -, 150, -, 30
Isaac Lewis, 35, 29, 1100, 75, 560
Abrm. Rodbaugh, 20, 61, 1000, 25, 300
Martin Teets, 15, -, 200, 15, 200
John Dix, 225, 381, 9000, 100, 1130
Peter Jackson, 33, 30, 600, 13, 170
Wm. Kiddy, 30, 120, 400, 23, 150
Jona Martin, 75, 65, 2600, 75, 500
C. Ciders, 12, -, 200, 13, 100
Ant. Teets, 58, 3, 1500, 100, 300
Able Clark, 6, -, 150, 25, 150
Wm. Allman, 11, -, 200, 13, 75
S. H. Nicholas, 10, -, 150, -, 25
Tho. Sutter, 8, -, 160, 12, 25
Tho. Bailey, 14, -, 180, 25, 70
Jac. Lorentz Jr., 180, 277, 9000, 80, 470
Thos. Kidd, 10, 8, 200, 23, 150
Saml. Mays, 8, -, 190, 12, 70
L. E. Price Sr., 8, -, 200, -, 20
Henly Eskin, 20, 63, 800, 25, 340
Abr. Allman, 60, 47, 1100, 22, 140
Armstead Queen, 60, 140, 2400, 60, 310
Richd. Altop, 17, -, 250, 25, 125
Jas. M. Romine, 40, 55, 1150, 40, 125
Robt. Rogers, 11, -, 200, 22, 40
J. M. McWhorter, 125, 207, 3500, 75, 475
B. Lawman, 35, 81, 800, 25, 125
F. M. Peters, 25, 56, 750, 25, 140
Josh. G. Peters, 23, 77, 650, 25, 120
J. S. Warner, 5, -, 100, -, 25
D. D. Casto, 30, 112, 1600, 75, 370
Wm. Warner, 128, 150, 3600, 50, 475
V. Hinkle Sr., 240, 125, 7450, 60, 1360
A. Post, 250, 100, 9000, 60, 1350

Jacob Waugh, 100, 60, 2400, 60, 600

Jer. Lance, 8, -, 160, -, -

David Neeley, 10, -, 200, 20, 150

Wm. Bargerhoff, 45, 30, 1400, 25, 80

Jonas Bargerhoff, 44, 30, 1300, 36, 250

Jno. Wentz, 50, 80, 1000, 22, 100

A. P. Faught, 22, 5, 1000, 85, 300

Alex. Riggs, 50, 75, 1250, 25, 380

W. G. Jennings, 20, -, 300, 60, 150

D. C. Napier, 15, 65, 600, 23, 150

R. A. Napier, 75, 215, 2200, 25, 125

Abyah Hinkle, 37, 12, 800, 100, 535

Wm. A. Johnson, 14, -, 140, 2, 100

Chs. Smith, 540, 200, 26000, 60, 4475

Henry Winemiller, 30, 65, 1000, 25, 240

Granville Post, 15, 125, 2100, 180, 250

N. H. Hannah, 200, 100, 6000, 100, 970

J. B. Shipman, 25, 129, 1200, 20, 200

Mic. Boyle, 17, 31, 500, 25, 150

Ashel Cutright, 35, 28, 1000, 10, 100

G. T. Gould, 140, 100, 4300, 65, 425

J. P. Piper, 50, 150, 1000, 75, 275

S. S. Black, 35, 90, 1800, 25, 210

G. D. Marple, 80, 45, 1000, 45, 310

W. O. Gould, 75, 25, 1500, 60, 450

A. Morgan, 100, 350, 3500, 20, 470

B. M. Pringle, 20, 130, 750, 12, 30

Isaac Cutright Jr., 80, 130, 2200, 28, 175

E. D. Rollins, 100, 100, 1600, 25, 200

V. Hinkle Jr., 30, 166, 1500, 25, 165

West Mills, 10, 20, 300, 25, 150

Jason Loomis, 100, 300, 5000, 60, 595

Lott Cutright, 100, 160, 4000, 25, 240

L. Phillips, 30, 45, 375, 25, 260

Geo. M. Talbott, 100, 140, 3500, 40, 510

Jac. Clark, 84, 30, 2300, 60, 600

B. F. Brown, 20, 1300, 1000, 22, 90

Martin Hinkle, 20, 55, 650, 25, 200

F. Phillips, 60, 40, 1250, 25, 200

Thamer Cutright, 20, -, 200, -, 30

Richd. Phillips, 30, 42, 1200, 85, 300

H. H. Lewis, 4, 46, 250, -, 35

Silas Roam, 10, 40, 150, 10, 60

B. Rohrbaugh, 110, 60, 3500, 25, 900

John Fultz, 10, 20, 260, 10, 120

Riley Reger, 150, 75, 4000, 60, 815

Jackson Calhoon, 30, 50, 1200, 25, 340

S. H. Bailey, 30, 117, 1600, 25, 200

S. H. Tate, 10, -, 200, -, 20

Hyre Brake, 20, 120, 1400, 20, 375

Philip Smith, 50, 90, 1700, 25, 300

David Morrisett, 10, 90, 800, 20, 30

Peter Cutright, 30, 33, 800, 30, 150

Goodman Reger, 80, 138, 3500, 100, 515

Geo. Allman, 252, 140, 8500, 150, 1060

J. N. Lorentz, 20, -, 300, 33, 275

Jacob Lorentz, 280, 1574, 21675, 150, 340

Geo. Stewart, 10, -, 150, 13, 45

V. J. Lorentz, 20, -, 400, 90, 500

Jno. Stewart, 10, -, 200, 10, 100

Geo. W. Lorentz, 200, 240, 13500, 200, 850

Tho. G. Sutter, 8, -, 160, -, 20

Nathan Allman, 5, -, 100, 20, 60

Henry Farrow, 15, -, 250, 35, 110

Howard Stewart, 10, -, 200, 22, 100

H. Winemiller, 5, -, 100, -, 20

Saml.Westfall, 30, -, 300, 23, 100

Peter Westfall, 100, 39, 2400, 20, 460

Geo. A. Westfall, 15, -, 300, 23, 80

Bevin Abbott, 40, 35, 900, 20, 328

Jas. Hersman, 50, 100, 1250, 20, 300

Martin Casto, 8, -, 98, -, 25

Nic. B. Linger, 12, -, 300, -, 30
J. Montgomery, 20, 75, 450, 75, 55
Geo. Herman, 75, 175, 1000, 25, 100
Nat. Moore, 150, 210, 4500, 120, 710
W. B. McCue, 80, 520, 4000, 30, 540
W. Linger, 50, 280, 3500, 30, 350
D. D. Smith, 60, 100, 1650, 40, 425
Johnson Smith, 10, -, 150, -, 120
Wm. Cutright Jr., 100, -, 100, 13, 110
Jona. Marsh, 25, 150, 700, 50, 130
B. Patterson, 40, 86, 2000, 45, 350
W. T. Higginbothem, 220, 125, 8625, 250, 1352
A. J. Hoffman, 15, 85, 600, 23, 75
Jas. Hoffman, 30, 267, 1400, 25, 265
Elmore Brake, 100, 100, 5000, 150, 340
Elijah Hyre, 45, 58, 2500, 60, 220
Cyrus Armstrong, 100, 105, 5250, 25, 285
Jacob Ables, 5, -, 100, 13, 85
Saml. Ables, 5, -, 100, -, 20
T. W. Tillman, 80, 50, 3750, 350, 700
M. Slaughter, 8, -, 120, -, 30
Geo. Brown, 67, 20, 2200, 20, 340
Abr. Shaver, 10, -, 200, 13, 120
R. K. Hocker, 35, 45, 2000, 75, 120
J. T. Cummings, 123, 70, 5400, 100, 900
J. R. Abott, 40, 20, 1500, 20, 250
Teter Lewis, 40, 18, 1200, 25, 200
Ant. Reger, 150, 160, 7350, 60, 880
F. A. Dowell, 10, -, 150, 10, 110
Noah Hyre, 125, 60, 4000, 50, 950
Jno. J. Reger, 220, 30, 5500, 100, 525
Mont. Reger, 20, -, 400, -, 400
H. Graves, 10, -, 100, 10, 75
Jas. Rucker, 5, -, 100, -, 25
Jas. E. Slaughter, 5, -, 100, -, 30
Wm. Sexton, 130, 116, 5000, 100, 810

Jas. M. Hoover, 8, -, 120, 10, 150
Saml. Gross, 40, 9, 500, 50, 215
Rebecca Liggett, 70, 160, 2500, 10, 190
A. Dowell, 40, 110, 1200, 3, 23
Jno. Morrison, 75, 50, 1250, 20, 400
Thos. D. Rucker, 6, 19, 250, -, 45
M. B. Hamner, 70, 147, 2170, 10, 330
Jacob Crise, 15, -, 300, 10, 250
W. G. Harlas, 50, 76, 1260, 60, 520
Geo. Cutright, 80, 30, 1750, 10, 500
Amos Cutright, -, -, -, 5, 130
J. W. Lemon, 80, 40, 2000, 20, 370
Jac. Crites Jr., 30, 79, 850, 25, 120
J. W. Abbott, 20, 10, 400, 15, 120
John Hyre, 30, 10, 900, 20, 300
Saml. Snider, 33, 100, 1000, 5, 100
Wm. Leedridge, 40, 59, 1500, 20, 310
W. S. Brady, 60, 40, 2000, 25, 300
Jas. Black, 30, 24, 850, 20, 300
J. H. Hodges, 5, -, 100, 25, 300
Jno. Hefner, 80, 26, 7500, 110, 600
Jno. H. Pritt, 40, 10, 1000, 30, 410
Ashley Gould, 250, -, 6250, 50, 1400
Dwight Gould, 128, 100, 5000, 40, 350
Elizabeth Hyre, 100, 100, 2500, 65, 600
J. T. Hyre, 10, -, 200, 25, 200
Benj. Gould, 150, 163, 4500, 40, 400
Cyntha Young, 95, 105, 2400, 20, 300
S. B. Young, 35, 50, 1200, 20, 305
Jno. Winemiller, 22, 23, 400, 25, 160
Goodman Simons, 35, 42, 650, 24, 170
John Simons, 75, 57, 1200, 25, 100
Elias Simons, 13, 87, 400, 25, 200
Noah Winemiller, 50, 69, 850, 25, 300
R. Holbert, 13, 25, 250, 12, 100
Jac. Duncan, 5, -, 100, 3, 30
Easter Jack, 200, 133, 6200, 75, 550

Henry Winemiller Sr., 100, 50, 3000, 25, 780

A. R. Jack, 12, 68, 1000, 25, 200

Wm. Henderson, 250, 250, 10000, 120, 1120

Jno. B. Henderson, 300, 300, 12000, 250, 1300

Henry Jones, 200, 83, 5600, 100, 1200

Tho. W. Vincent, 80, 59, 2100, 40, 630

Joseph Jones, 100, 51, 3000, 60, 738

Theo. Morgan, 145, 850, 4800, 210, 1100

Danl. Gould, 60, 60, 1800, 20, 510

Aaron Gould 120, 130, 6000, 50, 800

Thos. Brown, 8, -, 100, -, 50

E. Simons, 6, -, 120, -, 70

Elmore Hyre, 50, 50, 1200, 20, 275

N. M. Ferrell, 70, 67, 2200, 50, 435

H. M. Duglass, 10, -, 200, 2, 40

E. Leonard Jr., 700, 300, 26000, 260, 3600

S. A. Winemiller, -, -, -, -, 30

F. W. Sexton, 40, 65, 1300, 20, 325

J. S. Jack, 20, 92, 700, 15, 200

John Shobe, 25, -, 500, 15, 560

Jno. Shaver, 160, 73, 2500, 10, 100

Festus Young, 56, 76, 1200, 10, 330

J. A. Morgan, 300, 275, 8000, 150, 1500

Benj. Suddard, 10, -, 200, 3, 30

David Fairbun, 10, -, 150, 3, 35

W. A. Hosefloack, 8, -, 100, -, 35

A. L. Crites, 8, -, 100, -, 30

S. J. Rohrbaugh, 40, 135, 1500, 15, 150

Jno. Casto, 30, 22, 600, 3, 100

Annias Casto, 11, -, 150, 3, 100

Wm. Cutright, 70, 155, 2500, 100, 320

Jno. J. Burr, 600, 600, 18000, 150, 500

Danl. Bassell, 100, 130, 7000, 10, 1090

Job Simons, -, -, -, 3, 100

Sthiel Hinkle, 70, 5, 900, 50, 260

C. Cutright Sr., 20, -, 300, -, 50

C. O. Cutright, 8, -, 100, 4, 40

L. D. Cutright, 7, -, 80, 1, 100

Enoch Cutright, 175, 125, 3500, 40, 430

S. S. Laine, 63, 50, 1000, 10, 150

A. C. Pringle, 16, 34, 400, 5, 80

Joel Pringle, 10, 15, 200, 3, 30

Jno. Cutright Jr., 5, -, 60, 3, 30

Jno. H. Boyle, 16, 56, 600, 3, 40

Theo. Cutright, 90, 149, 2000, 15, 175

Dexter Cutright, 8, -, 100, 5, 150

Jac. Cutright Sr., 79, 50, 2500, 30, 165

Abram Strader, 300, 335, 10500, 120, 1250

W. L. Coldrider, 20, 310, 1300, 15, 160

Geo. Carper, 135, 100, 5000, 100, 875

John Miller, 22, -, 300, 10, 120

Asa Strader, 90, 46, 1200, 20, 360

Henry Owens Jr., 25, -, 250, 20, 140

H. C. Middleton, 20, 23, 5000, 75, 245

John Owens, 20, -, 300, 20, 175

Robt. Coyner Sr., 11, -, 1100, 75, 220

James Tagg, 10, 40, 150, 15, 75

Ira Graves, 10, -, 200, 15, 100

R. H. Parrick, 18, -, 250, -, 25

Geo. W. Berlin, 200, 2250, 14200, 100, 225

Jno. A. Foster, 30, 120, 800, 5, 160

E. J. Coldrider, 13, -, 2000, 20, 30

Geo. Warner, 20, -, 400, 25, 275

Geo. D. White, 80, 180, 4000, 300, 500

Elias Heavener, 100, 40, 3500, 120, 535

Henry Simpson, 2, -, 500, 10, 150

Lorenzo Pumphry, 10, -, 200, -, 35

C. D. Trimble, 75, 231, 7000, 100, 440

Nat. Cutright, 184, 80, 3200, 180, 510

N. Farnsworth, 27, -, 2500, 20, 75

T. J. Farnsworth, 115, 300, 6500, 10, 160

Levi Leonard, 57, 23, 6100, 140, 840

F. Gilmer & Wife, 5, -, 1600, 5, 35

W. D. Farnsworth, 10, 130, 1800, 6, 250

Geo. Bastable, 381, 62, 20210, 120, 2500

S. Hoffman, 40, 60, 1000, 20, 205

Danl. Carper, 600, 300, 18000, 125, 2750

M. T. Humphry, 32, 70, 1020, 25, 200

J. R. Hodges, 35, 35, 70, 120, 340

M__ Rohrbaugh, 160, 40, 4000, 100, 615

N. B. Foster, 15, -, 3200, 15, 80

N. B. Jackson, 100, 80, 2850, 20, 400

Amos. Grubb, 60, 25, 800, 15, 100

N. S. Holland, 20, -, 400, 15, 200

T. B. Rothwell, 20, -, 300, 15, 150

Jno. Linch, 50, 59, 1400, 40, 380

C. B. Mayo, 95, 228, 2200, 20, 475

C. W. McNulty, 75, 50, 3600, 120, 590

T. P. Chipps, 16, -, 600, 15, 110

D. E. Coyner, 70, 30, 1200, 100, 700

Judson Hinkle, 60, 48, 1500, 10, 100

Joseph Coyner, 60, 44, 3000, 75, 120

Jno. Dean Jr., 15, -, 200, 10, 100

Josiah Leggett, 80, 80, 4000, 50, 250

J. G. Sutton, 10, -, 300, 5, 75

A. Thompson, 90, 98, 3000, 220, 625

Enoch Gibson, 120, 380, 5500, 120, 660

W. J. Dean, 45, 24, 280, 15, 225

Geo. W. Owens, 2, -, 600, 130, 210

Henry Owens, 160, 240, 4400, 50, 100

J. D. Fretwell, 5, -, -, 3, 30

Arc. Pumphry, 6, -, -, 2, 35

J. B. Shreve, 100, 1350, 6500, 240, 535

Joseph Miller, 40, -, 800, 20, 320

A. J. Wood, 65, 122, 1400, 30, 270

J. A. Watson, 20, -, 400, 5, 45

Elijah Harper, 17, 50, 900, 15, 190

Ab. Shreve, 114, 180, 2900, 20, 400

Lear Dean, 47, 75, 900, 10, 175

M. E. Shreve, 20, 30, 600, 35, 130

Adam Boyer, 25, 75, 550, -, 100

S___ Simons, 21, 32, 400, 5, 100

Wm. Kusling, 40, 13, 550, 15, 225

John Kusling, 74, 25, 900, 20, 175

M. Lewis, 10, -, 100, 5, 20

Asa Fornash, 25, 25, 600, 10, 75

Jackson Ward, 20, 84, 1000, 15, 90

John Grimm, 40, 33, 550, 15, 190

Marshall Dean, 50, 200, 950, 10, 230

Ant. Pifer, 40, 10, 400, 15, 235

Isaac Kusling, 70, 60, 1800, 20, 250

Jas. Kusling, 70, 50, 1600, 20, 320

Jos. Bosley, 35, 250, 1000, 60, 120

John Shreve, 40, 150, 800, 20, 250

Wm. Dolin, 20, 30, 300, 5, 100

A. J. Harris, 70, 430, 1000, 15, 130

P. B. Williams, 16, 114, 500, 10, 215

M. Welsh, 6, 94, 300, -, 40

Jno. Ryan, 30, 70, 500, 15, 100

J. W. Johnson, 3, -, 50, 20, 190

Joel A. Foster, 40, 15, 550, 50, 140

Chas. D. Hess, 40, 110, 1200, 10, 180

Jas. McDermot, 60, 240, 1000, 5, 200

Pat. Owens, 8, 42, 150, 5, 25

Patrick Ferren, 7, -, 90, 3, 95

Jno. B. Hilley, 100, 130, 5000, 200, 750

Thos. Leonard, 12, 88, 350, 5, 30

Simon Strader, 50, 120, 1000, 60, 340

N. B. Wamsley, 7, -, 200, 250, 335

B. C. Horntack, 30, 370, 1600, 20, 165

W. S. King, 21, 20, 800, 30, 200

Jos. H. Wicks, 12, 68, 250, 5, 35

R. L. Duly, 20, 150, 400, 15, 145
J. J. Thacker, 35, 89, 1000, 20, 125
Jas. Trusler, 33, 33, 450, 5, 35
J. Macdearmott, 75, 255, 2250, 20, 450
T. Harris, 25, -, 300, 15, 80
Danl. Marting, 40, 160, 600, 10, 200
John Boothe, 9, -, 100, 23, 125
Wm. Smith, 10, -, 150, 5, 60
Patrick Gadolee (Godolee), 10, 190, 300, 2, 35
T. Cunningham, 50, 250, 1000, 20, 200
A. Rodebaugh, 80, 316, 4000, 100, 500
A. J. Stemple, 10, -, 100, 13, 130
Wm. Wagner, 30, 120, 300, 25, 260
Andrew Boyle, 30, 100, 300, 15, 125
Thos. Brady, 10, 115, 300, 13, 110
Lewis Zeckle, 10, -, 100, 3, 40
E. J. Obrian, 100, 1400, 3000, 75, 100
Wm. Dunnington, 30, 50, 160, 25, 200
M. Hoover, 15, 100, 200, 12, 150
S. George, 10, -, 100, 5, 10
N. Heavener, 47, 55, 600, 20, 165
Arthur Wood, 12, -, 140, 15, 100
Nic. Dean, 45, 48,700, 20, 250
V. Dickerson, 50, 110, 1250, 20, 230
Wm. Griffith, 35, 65, 600, 15, 200
T. Marshall, 50, 158, 1600, 80, 310
Pat. Burke, 20, 240, 500, 10, 200
G. L. Mayo, 40, 120, 1000, 25, 305
Martin Murry, 10, 50, 240, 3, 100
T. Dolin, 8, 92, 300, 2, 50
Saml. Hiner, 15, 138, 400, 100, 275
Jac. Paugh, 50, 135, 1000, 20, 210
Isaac White, 125, 375, 5000, 250, 350
Stephen Danson, 20, -, 300, 25, 175
Chrs. Simons, 10, -, 200, 25, 175
L. Sandridge, 40, 74, 1000, 20, 220
Luther Sandridge, 10, -, 100, 20, 220
Jos. Houser, 150, 146, 3500, 200, 555

Jac. Neff, 30, 75, 350, 15, 160
Isaac Moreland, 10, 120, 350, 25, 350
Isaac Strader, 100, 205, 2100, 20, 290
Benj. Strader, 20, -, 300, 10, 250
Nich. Michaels, 10, -, 200, 15, 160
Rachel Rollens, 15, -, 225, 15, 185
Wat. Westfall, 40, 60, 1200, 20, 100
Solo. Reese, 50, 25, 1000, 20, 155
David Reese, 25, 15, 450, 5, 60
M. Westfall, 75, 133, 1650, 20, 200
T. Wilfong, 24, 4, 160, 12, 45
Ebrm. Cutright, 75, 25, 1000, 30, 500
Nelson Jones, 40, 56, 750, 15, 350
Geo. Clark, 100, 125, 2600, 75, 120
Saml. Boyer, 5, -, 100, 3, 40
Geo. W. Burner, 86, 160, 3000, 100, 460
William Bean, 20, 80, 600, 15, 200
T. A. Norvell, 4, -, 50, 3, 37
John Tenney, 10, -, 60, 3, 35
Nich. Owens Jr., 25, 35, 300, 15, 110
Jas. Brian, 10, -, 100, -, 10
Jas. M. Black, 60, 40, 900, 10, 200
M. V. Black, 7, -, 88, 3, 120
Peter Tenney, 60, 50, 1000, 25, 400
Nic. Owens Sr., 40, 180, 1000, 20, 100
William Moody, 12, 22, 300, 15, 170
Abr. Owens, 100, 278, 2800, 10, 490
Peter Barb, 10, -, 120, 15, 120
Silas Barb, 75, 60, 1600, 150, 575
W. L. Barb, 10, -, 150, 25, 200
J. W. Currants, 20, 140, 400, 15, 180
Lydia Tenney, 50, 102, 600, 15, 200
F. S. Kettle, 5, -, 50, 2, 50
Peter Tenney Jr., 15, 11, 200, 14, 300
E. Montgomery, 6, -, 75, 2, 30
Jos. Dean, 40, 10, 400, 15, 175
E. C. Rollens, 25, 55, 400, 13, 80
Isaac Wamsley, 25, -, 250, 10, 165
Timothy Mexon, 15, -, 200, 15, 80
Eli Wilfong, 25, -, 250, 15, 80

Jno. Slaughter, 10, -, 100, 13, 80

Benj. Bassell, 400, 95, 15000, 250, 3510

W. Summers, 465, 157, 15000, 150, 1750

Jno. Baily, 10, -, 150, 3, 50

T. F. Payne, 35, -, 350, 100, 240

Nimrod Reger, 223, 150, 9375, 100, 630

H. A. Fury, 7, -, 175, 3, 50

Jno. B. Brake, 150, 23, 3500, 75, 583

Jac. J. Brake, 25, 8, 500, 15, 40

Henry Wilfong, 20, 31, 500, 2, 30

Co_. Cutright, 12, 16, 120, 13, 85

Filo Tenney, 10, -, 200, 2, 100

Jas. Gooden, 20, 170, 500, 20, 100

Geo. S. Groves, 20, 31, 500, 100, 210

Jno. L. Boggess, 30, 130, 600, 50, 250

John Dean, 100, 35, 3350, 100, 620

Val. Strader, 320, 290, 12100, 100, 1150

G. W. Ratliff, 28, -, 350, 10, 35

D. J. Carper, 125, 75, 5000, 100, 1750

Martin Burr, 8, -, 100, 3, 25

A. C. Shreve, 47, 90, 1200, 10, 265

Elizabeth Foster, 60, 91, 900, 25, 238

Solo. Day, 130, 188, 3000, 200, 630

H. A. Fultz, 14, -, 180, 5, 45

Jesse P. Sharp, 15, 137, 600, 15, 45

Jacob Snell, 100, 336, 2000, 100, 325

M. Newcomb, 15, 25, 300, 15, 55

Isaac Strader Jr., 37, 115, 900, 20, 230

Nathan Liggett, 60, 90, 1250, 30, 300

Elijah Gooden, 30, 70, 900, 15, 105

W. B. Gooden, 70, 130, 1800, 30, 125

Darcus Tenney, 26, 50, 1000, 10, 210

A. R. Tenney, 25, 53, 450, 10, 85

Marshall Tenney, 12, 38, 400, 5, 40

Wm. Tenney, 5, 45, 300, 3, 35

Geo. W. Tenney, 40, 112, 150, 20, 260

Js. C. Tallman, 14, -, 186, 15, 30

Benna Tallman, 100, 412, 2500, 75, 400

R. B. Tallman, 10, -, 150, 5, 100

W. B. Tallman, 26, 125, 1000, 25, 570

Able Wilfong, 12, -, 150, 5, 100

Ed Grims, 11, -, 130, 23, 130

Agr. Osban, 20, 8, 300, 15, 110

Susanna Gooden, 50, 100, 500, 5, 115

Geo. T. Herenden, 50, 250, 2000, 60, 550

R. C. Wingfield, 68, 400, 3000, 75, 350

Jno. R. Wingfield, 25, 109, 800, 15, 120

Nelson Wingfield, 8, -, 160, 2, 40

H. Barrackman, 30, 100, 350, 15, 80

Jac. Snider, 65, 235, 1800, 75, 435

Jno. L. Tenney, 10, 190, 300, 15, 150

S. McCann, 10, -, 150, 25, 325

S. B. McCann, 4, -, 60, 15, 110

Jno. House, 40, 63, 900, 15, 135

Henry Bean, 45, 205, 1000, 10, 290

Jac. Cutright Jr., 20, 30, 300, 15, 100

Andrew Bean, 20, 30, 350, 10, 205

Henry Bosley, 30, 70, 700, 15, 175

Joseph Gould, 40, 260, 1200, 15, 250

David Phillips, 40, 40, 700, 50, 250

Jas. Gladwell, 40, 35, 450, 15, 150

Jesse Lemon, 40, 60, 600, 25, 200

Noah Westfall, 16, -, 200, 3, 90

J. W. Simons, 40, 400, 1600, 50, 150

Jas. D. Simons, 10, -, 200, 13, 100

Abr. Crites, 15, -, 200, 15, 150

Leonard Crites, 15, -, 200, 15, 150

Jac. L. Crites, 80, 920, 5000, 50, 465

Jas. Esken, 10, -, 150, 15, 110

Josiah Martin, 15, -, 180, 15, 85

Thos. Shaw, 10, -, 150, 15, 110

Peter Hoffman, 5, 195, 200, 50, 120

Levi Simons, 5, 45, 100, 13, 95
Wm. Parker, 8, 92, 200, 15, 85
Wm. Harris, 6, -, 120, 5, 30
Peter Johnson, 15, 185, 600, 15, 150
Levi Paugh, 25, 125, 600, 24, 260
Danl. Phipps, 35, 45, 220, 20, 330
Perry Johnson, 10, 192, 600, 20, 80
Jesse Johnson, 12, 136, 600, 20, 1809
Albert Carter, 14, 90, 600, 5, 70
Geo. W. Gladwell, 5,-, 60, 13, 30
Wm. Smallridge, 110, 170, 3500, 40, 620
Geo. Andrews, 30, -, 400, 40, 250
Lewis Lunsford, 20, 55, 1600, 25, 130
Elijah Brain, 20, 390, 1100, 20, 360
Geo. Talbott, 5, -, 100, 15, 80
Wm. Reed, 50, 50, 1500, 15, 200
Chap. McCoy, 35, 115, 1500, 25, 200
Delia Perry, 8, 300, 300, 15, 100
Elzy Haddox, 5, -, 75, 3, 30
Wm. Phillips, 40, 10, 600, 20, 220
Horace Phillips, 40, 60, 800, 10, 175
Eber. Phillips, 40, 73, 900, 15, 230
Jno. Phillips, 20, 113, 700, 10, 125
Silvester Phillips, 30, 160, 800, 15, 110
John P. Phillips, 6, 40, 300, 15, 135
Elzy Vancamp, 30, 100, 400, 20, 140
Uriah Phillips, 40, 95, 1000l, 20, 240
Jno. S. Thomas, 70, 30, 1100, 100, 625
Edwin Perry, 15, -, 150, 5, 50
E. G. Burr, 500, 890, 23500, 190, 2440
B. Lentz, 7, -, 140, 5, 40
A. J. Hinkle, 75, 62, 3000, 15, 570
Martin Reger, 70, 20, 2500, 120, 930
Marshal Hyre, 15, 19, 600, 20, 140
Barbery Reger, 40, -, 1200, 25, 350
S. B. Fretwell, 60, 140, 2000, 60, 220
Jno. D. Hyre, 260, 135, 9450, 150, 1700

Sol. Cutright, 10, -, 150, 5, 50
Leml. Brake, 80, 80, 1750, 80, 500
Randolph See, 135, 35, 2800, 70, 770
S. B. Winemiller, 10, -, 200, 5, 40
Adolphus Brooks, 10, -, 200, 50, 130
Washington Sexton, 12,-, 200, 100, 150
Jas. Sexton, 90, 73, 2500, 100, 500
Saml. Smallridge, 10, -, 150, 15, 130
H. Armstrong, 100, 230, 5000, 150, 760
Jas. Blagg, 18, -, 450, 5, 75
Jas. Smallridge, 60, 40, 1600, 30, 350
Stewart Bennett, 200, 28, 5000, 250, 1055
J. M. Smith, 20, -, 400, 100, 750
D. Bennett, 200, 688, 7900, 50, 1250
Wm. Bennett, 40, 460, 1200, 20, 225
Jno. Duglass, 160, 168, 4500, 120, 980
A. P. Rusmisell, 36, 5, 1000, 75, 310
Sally Curry, 80, 45, 2200, 75, 620
Jas. A. Campbell, 25, 100, 1500, 20, 85
Mary Curry, 50, 5, 500, 10, 170
James Curry, 60, 119, 1800, 50, 350
Leo. Nicholas, 30, 35, 800, 20, 160
Jno. A. McDowell, 25, 27, 750, 50, 145
W. E. McDowell, 40, 63, 1300, 50, 150
John Smith, 90, 153, 3000, 100, 510
Noah Smith, 15, 227, 900, 20, 170
D. B. Smith, 25, 75, 800, 20, 240
Robert Curry, 75, 125, 2400, 20, 200
Jas. T. Hull, 9, 402, 6000, 40, 600
Saml. Wilson, 225, 753, 9000, 100,480
Jno. W. Wilson, 30,-, 500, 40, 260
Jno. Armstrong, 120, -, 2400, 100, 220
Jno. W. Armstrong, 50, 37, 1350, 50, 550
_. M. Bennett, 5, -, 100, 20, 130

S. T. Talbott, 150, 2850, 15000, 100, 740

Elizabeth Lanham, 10, -, 200, 5, 80

D. J. Talbott, 140, 235, 4200, 50, 975

Jas. Duglass, 20, 30, 500, 20, 140

Jared Armstrong, 50, 96, 2300, 40, 335

R. H. Townsend, 100, 50, 3500, 150, 850

J. A. Wood, 5, -, 50, -, 30

C. W. Townsend, 60, 90, 3000, 25, 450

Robert Johns, 60, 150, 2100, 75, 325

Davis K. Johns, 10, -, 150, 15, 100

J. F. Friel, 38, 70, 1000, 20, 150

Isaac Fleming, 40, 12, 500, 20, 170

G. A. Fleming, 20, 18, 350, 20, 110

John Strader, 110, 100, 4000, 100, 830

Danl. Waggy, 40, 75, 900, 20, 215

Andrew Collins, 35, 65, 800, 20, 215

Jno. Collins, 25, 195, 1300, 20, 130

Peter Harper, 12, 48, 300, 20, 120

Jef. C. Vincent, 10, -, 150, 23, 110

W. B. Hull, 40, 160, 1600, 50, 535

Jas. S. Wilson, 6, -, 120, 5, 45

Jos. Crawford, 35, 75, 700, 20, 255

Obd. Crawford, 35, 75, 750, 20, 230

Val. Siron, 30, 32, 350, 15, 150

W. N. Childress, 20, 80, 800, 20, 140

W. H. McClain, 15, -, 200, 20, 230

Peter Gum, 12, 88, 500, 15, 130

Jno. D. Loudin, 45, 165, 1400, 20, 250

N. C. Brake, 45, 105, 800, 20, 435

Amos Sample, 31, 250, 1500, 20, 150

Jas. Brown, 80, 310, 1500, 20, 450

A. Cunningham, 30, 106, 1100, 35, 225

Thos. Cunningham, 46, 200, 1800, 30, 160

Jno. A. Cunningham, 20, 80, 800, 20, 160

Wm. Fleming, 20, 80, 800, 20, 300

Jas. Cockran, 25,-, 500, 30, 150

Ad. Harper, 18, 147, 600, 25, 140

Saml. Loudin, 25, 100, 500, 20, 120

J. H. Shaver, 50, 1050, 2220, 25, 450

Saml. Marley, 20, -, 250, 15, 100

Wm. Fiddler, 90, 110, 2500, 150, 450

L. F. Corbett, 40, 1060, 3000, 100, 480

Tho. P. Despre, 30, 170, 600, 120, 250

B. F. Clarkson, 25, 52, 250, 20, 335

Jno. Helmick, 10, -, 150, 10, 120

Geo. Eagle, 65, 135, 1200, 20, 385

A. Beverage, 68, 207, 1500, 25, 180

Thos. McQuain, 50, 150, 1400, 20, 340

Agnes Brake, 70, 90, 1600, 50, 450

A. B. Vincent, 10, -, 200, 10, 50

L. W. Ferrell, 20, 32, 500, 50, 30

A. W. Gum, 30, 30, 720, 20, 140

J. J. Brake, 60, 70, 1560, 25, 450

N. Heavener, 40, 60, 1200, 30, 300

John Baker, 10, -, 280, 80, 60

C. S. Haynes, 80, 220, 3000, 40, 275

Robt. McAvoy, 50, 75, 2000, 75, 390

Jas. McAvoy, 50, 75, 2000, 75, 440

H. M. Hamner, 10, 75, 600, 20, 90

Chet Morgan, 30, 60, 2500, 120, 210

Lyman Young, 25, 25, 750, 25, 95

Jona. Potts, 19, -, 380, 80, 400

Gilbert Young, 40, 56, 1500, 25, 255

Edwin Phillips, 114, 100, 200, 25, 260

Almanders Young, 10, 60, 280, 10, 30

Jac. Smith, 15, -, 180, 20, 120

Jas. Duglass, 20, 30, 600, 25, 170

M. A. Darnell, 38, 60, 1200, 35, 165

B. J. Miles, 60, 260, 3700, 60, 260

H. Perry, 20, 30, 650, 20, 100

J. W. Riggleman, 15, -, 300, 25, 15

T. C. Hall, 25, 26, 650, 25, 155

H. McKissick, 10, -, 150, 10, 70

Isaac Pritt, 12, -, 180, 10, 50

W. M. Haymond, 80, 1040, 4400, 20, 270

Martha Griffin, 7, -, 100, 2, 60
Wm. Pritt, 20, 30, 650, 20, 150
Hugh Agan (Agar), 26, 280, 1000, 20, 230
M. Corbitt, 20, 206, 1700, 20, 130
Susan Rakes, 20, 80, 300, 10, 65
Edwin Hyre, 30, 200, 1000, 20, 550
Jas. Pritt, 26, 214, 950, 25, 240
Geo. S. Riffle, 50, 369, 1600, 20, 150
Josiah Bennett, 15, 27, 175, 15, 155
Lewis Collins, 25, 150, 525, 15, 175
J. S. Strader, 40, -, 400, 20, 330
Salathel Strader, 15, 100, 350, 15, 100
T. O. Slaton, 10, -, 150, 10, 66
Geo. Lake, 20, -, 300, 20, 460
Elizabeth Green, 8, 67, 150, 5, 40
Jotham Rice, 70, 58, 6000, 80, 640
Jas. Long, 40, 72, 650, 20, 145
Wm. McCann, 5, -, 100, 5, 35
John Teter, 200, 1600, 9000, 150, 1750
Jesse Nixon, 30, 170, 1000, 20, 225
S. Martiney, 10, 190, 600, 20, 250
Earl E. Young, 25, 87, 900, 15, 140
Cornelius Clark, 20, 155, 1000, 20, 140
D. H. Morrison, 20, 142, 800, 15, 60
E. Bartlett, 60, 52, 1250, 50, 340
R. Bartlett, 30, 71, 1100, 15, 250

Ed. Young, 10, 40, 380, 15, 120
Tho. Gawthrop, 30, 270, 1500, 50, 300
Saml. Allman, 20, 50, 375, 20, 180
B. Guyer, 20, 105, 800, 20, 150
Mary Guyer, 20, 105, 750, 10, 170
Jas. Donnelley, 50, 80, 1750, 80, 500
J. J. Vincent, 110, 50, 2000, 80, 550
Peter Smith, 25, 115, 1650, 25, 270
E. & W. Townsend, 200, 282, 11000, 465, 1835
Maj. Tharp, 110, 150, 1000, 20, 120
Jno. J. Lemon, 10, 90, 300, 10, 90
Simon Casto, 30, 70, 500, 20, 170
Seneca Norvell, 8,-, 160, 5, 50
H. T. Garter, 25, 75, 800, 20, 140
Abr. Cutright, 20, 30, 500, 15, 140
Calvin Cutright, 10, 27, 250, 10, 30
Saml. Eakle, 30, 200, 3000, 60, 240
Christ. Eagle, 35, 155, 1500, 25, 130
M. J. Eakle, 15, -, 300, 15, 125
Geo. Bookins, 12, 338, 1000, 15, 110
Archd. Hinkle, 300, 165, 10000, 150, 2030
Wm. Harris, 20, -, 200, 20, 110
David Moss, 15, 45, 600, 80, 400
Chas. Demoss, 30, 106, 1000, 25, 100
Wm. Heavener, 13, 15, 1200, 25, 150

Wayne County, West Virginia
1860 Agricultural Census

The University of North Carolina at Chapel Hill filmed the 1860 agricultural census for Wayne County from originals at the West Virginia State Archives under a grant from the National Science Foundation in 1963.

Columns 1, 2, 3, 4, 5, and 13 represent the following information on the census:
1. Name of Owner, Agent or Manager of Farm
2. Acres of Improved Land
3. Acres of Unimproved Land
4. Cash Value of the Farm
5. Value of Farming Implements and Machinery
13. Value of Livestock

Some parts of this county were very faint and difficult to read.

Hugh Bowen, 210, 1835, 1200, 15, 302
Hiram Butterfield (Butterfard), 95, 104, 2000, 40, 400
Joel Caylor, 30, 45, 300, 10, 319
Samuel Booth, 70, 130, 2000, 10, 300
Jameson Ferguson, 15, 105, 400, 8, 150
Levi Morris, 75, 150, 3000, 15, 650
Burwell Spurlock, 90, 660, 8000, 100, 700
Benjamin Dean, 20, 40, 300, 10, 114
Thurston Spurlock, 60, 360, 400, 200, 488
Wm. Smith, 14, 14, 420, 10, 103
Burwell Booth, 40, 60, 300, 8, 105
Wm. Morris, 100, 800, 8000, 30, 285
Calvary Adkins, 12, 125, 500, 6, 200
Thos. Harrison, 30, 60, 1000, 60, 300
Andrew Adkins, 35, 320, 200, 40, 300
Morgan Garrett, 25, 190, 1200, 75, 250
John R. Chapman, 17, 17, 425, 4, 140
Isah Bloss Jr., 25, 335, 2000, 150, 244

Jesse Spurlock, 125, 700, 6000, 50, 400
Stephen Spurlock, 40, 520, 1600, 20, 337
Rebecca Spurlock, 50, 350, 3000, 100, 250
Jacob Plybon, 20, 84, 800, 30, 140
John A. Beckner, 20, 94, 250, 120, 365
Christopher Keysor, 75, 250, 200, 5, 181
John F. Barber, 20, 45, 800, 40, 293
Saml. A. G. Mcginis, 55, 262, 2000, 200, 334
Harvy Dunkle, 21, 1400, 420, 30, 300
John Dunkle, 30, 70, 1000, 5, 300
Hiram Bloss, 130, 80, 7000, 100, 765
Peter Angle, 30, 130, 800, 10, 175
Henry Hollenback, 70, 20, 300, 40, 275
Stephen Beckner, 18, 90, 90, 4, 88
Moses Rigg, 25, 75, 350, 3, 104
Wm. Rutherford, 35, 75, 700, 3, 112
John E. Paul, 50, 40, 450, 5, 202
Robert A. Ward, 40, 250, 750, 13, 218

Henry Luther, 70, 330, 4000, 50, 185
Charles R. Perdue, 20, -, 150, 4, 34
Rufus M. Luther, 22, -, 150, 5, 35
Merideth Allen, 35, -, 700, 3, 300
Joel Barber, 30, 70, 600, 5, 175
Morris Newman, 65, 50, 2000, 50, 200
Greenville Newman, 37, 200, 1700, 60, 223
Joseph Newman, 45, 110, 2500, 50, 305
John Newman, 27, 123, 1200, 25, 175
Robert C. Smith, 25, 50, 1000, 5, 300
James Barber, 18, 175, 1500, 10, 244
John Bloss, 91, 495, 2500, 200, 625
Jefferson Booth, 30, 60, 400, 6, 200
Benj. Garrett, 70, 500, 3000, 40, 217
John N. Smith, 100, 160, 3000, 150, 391
Burwell Booth, 15, 217, 1500, 5, 125
William Ferguson, 65, 300, 2500, 75, 293
Jefferson B. Bowen, 120, 200, 3700, 75, 879
Charles W. Ferguson, 120, 2100, 6625, 187, 770
Jno. Blankenship, 125, 700, 5000, 40, 300
Wm. H. Smith, 16, -, 320, 4, 160
Morris Booton, 15, 400, 5000, 75, 450
Frank Bowen, 15, 250, 50, 9, 188
Calander Spurlock, 50, 150, 1500, 35, 470
David Dick, 14, 80, 800, 8, 293
Noah Adkins, 35, 90, 1000, 7, 140
Dyke Bowen, 22, 129, 1000, 5, 134
Purl Ferguson, 8, 109, 400, 5, 103
Patrick Hensley, 65, 270, 2500, 200, 438
Lucy McComas, 50, 110, 2000, 20, 259
Daniel Lawhorn, 20, 30, 400, 7, 146

Saml. Blankenship, 17, 89, 600, 5, 73
R. P. Drown, 50, 110, 2000, 5, 140
Solomon Hensley, 40, 117, 1200, 30, 360
Mary Davis, 40, 120, 1100, 30, 300
James H. Davis, 12, 150, 600, 10, 125
Hugh B. Adkins, 40, 167, 1300, 10, 326
Daniel Davis, 125, 275, 3500, 100, 826
Ann F. Bowen, 70, 100, 1500, 300, 338
James P. Keysor, 40, 110, 1750, 100, 214
Hiram Cremeans, 35, 153, 612, 10, 175
Andrew Adkins, 60, 465, 5000, 50, 618
William Bush, 40, 503, 1000, 50, 98
Isaa Bloss, 120, 580, 3615, 200, 1121
James C. Bowen, 100, 730, 3000, 125, 553
Wesly Adkins, 8, 60, 160, 55, 110
Wm. Adkins, 60, 657, 3000, 50, 266
Isaac Adkins, 40, 42, 150, 4, 71
Joseph Adkins, 50, 650, 3000, 20, 635
John Adkins, 50, 650, 3000, 50, 342
Alexander Adkins, 45, 130, 1000, 10, 362
Hezekiah Adkins, 22, 30, 500, 20, 195
_. L. Adkins, 27, 100, 800, 25, 334
Hezekiah Adkins, 35, 150, 1000, 30, 230
Wm. Biens, 20, 60, 400, 6, 150
Frederick Morison, 14, 136, 500, 5, 52
Thomas J. Adkins, 14, 130, 1000, 3, 41
Elijah M. Adkins, 26, 274, 1500, 5, 115
Everward Adkins, 10, -, 200, 2, 80

Bengamin Childress, 60, 800, 1500, 10, 640
Vinson Lucus, 35, 40, 300, 5, 100
Sylvester Adkins, 75, 20, 1000, 20, 386
Raben Mays, 40, 125, 1000, 10, 275
___ly Adkins, 50, 450, 1600, 5, 175
Isom Adkins, 40, 140, 1200, 10, 204
Jacob Adkins, 20, 265, 800, 8, 138
B. T. T. Adkins, 15, 118, 1000, 12, 125
Delpha Adkins, 30, 60, 800, 5, 230
Lisander Adkins, 60, 540, 2000, 10, 419
Calvin Adkins, 63, 500, 2000, 5, 380
Pleasant Bartrum, 20, 100, 1000, 4, 239
Camford Adkins, 25, 60, 400, 5, 124
Ira Gilkison, 25, 100, 600, 5, 340
Thomas Gilkison, 30, 15, 300, 10, 173
L__ B. Adkins, 18, 50, 7000, 1, 233
Jones Adkins, 15, -, 300, 4, 250
Robert Ross, 14, -, 280, 100, 75
Charles Adkins, 71, 1122, 3086, 10, 859
Rease W. Elkins, 46, 250, 1500, 5, 98
Miles Elkins, 8, 9, 250, 8, 85
Thomas Parsons, 50, 125, 800, 5, 367
William Eplin, 15, 630, 600, 10, 84
Hiram Adkins, 50, 480, 2100, 5, 292
William Haney, 50, 600, 200, 30, 175
B. D. Massey, 65, 235, 1400, 20, 195
James M. Clay, 50, 120, 800, 5, 121
Edward Parsons, 25, 75, 800, 5, 150
Charles Adkins, 12, 288, 1000, 10, 146
John Price, 45, 195, 700, 15, 250
James Mills, 30, 90, 400, 10, 200
Oliver Mills, 30, 95, 400, 12, 30
Thomas Mills, 75, 164, 100, 80, 290
_. M. Morrison, 10, 40, 100, 5, 27

James R. Morrison, 43, 159, 800, 40, 275
Asa Morrison, 8, 42, 200, 4, 125
Norman Morison, 20, 115, 200, 5, 120
Morris Gilkison, 60, 1300, 3000, 20, 500
John Gilkison, 25, 225, 800, 20, 250
William J. Adkins, 75, 700, 3000, 50, 450
A. J. Crockett, 25, 135, 600, 5, 489
Edward J. Ferguson, 12, 369, 1500, 7, 179
Maxwell Bradshaw, 12, 60, 300, 7, 120
Elihu Luther, 18, 136, 600, 8, 101
Andrew Samon, 25, 150, 100, 20, 219
John E. Smith, 20, 50, 700, 20, 110
C. J. Ballenger, 25, 100, 600, 10, 196
John Piles, 70, 90, 1200, 15, 450
Jamison Booth, 50, 75, 1000, 6, 314
James W. Smith, 17, 83, 200, 4, 52
Wesly Booth, 80, 38, 800, 25, 295
William J. Wilkinson, 100, 220, 2500, 150, 610
Isaac Piles, 60, 92, 700, 100, 209
D. C. Canada, 80, 45, 2000, 10, 543
John R. Miller, 90, 200, 2500, 100, 615
Harrison Adkins, 40, 210, 1500, 3, 136
Washington Ferguson, 40, 160, 700, 20, 435
Hezekiah Neal, 40, 160, 600, 100, 137
Harvy Shannon, 50, 130, 700, 15, 236
Hiram Lott, 15, 120, 500, 4, 82
Thomas V. Barkish, 40, 85, 1000, 10, 203
Charles Parks, 45, 120, 700, 5, 73
Wm. Thompson, 25, 75, 700, 6, 104
Wesly Peery, 9, 91, 500, 4, 30
James Thompson, 20, 91, 800, 5, 147
John Duke, 25, 100, 500, 5, 180

Thomas Cyrus, 80, 25, 1000, 3, 150
M. D. L. Burns, 18, 85, 600, 3, 34
Ruban Walker, 120, 200, 4500, 225, 600
James Peters, 20, -, 1000, 10, 369
William C. Williams, 20, -, 1000, 10, 225
Hiram Frampton, 320, 100, 21000, 280, 970
James H. Ward, 11, -, 150, 3, 114
John L. Walker, 30, -, 200, 3, 149
Robert Morgan, 13, 100, 800, 3, 70
James Casey, 26, 74, 800, 3, 70
William Bond, 25, -, 250, 3, 108
Jesse Ward, 18, 100, 400, 63, 300
William Stewart, 100, 50, 2400, 150, 495
Wm. Casey, 20, -, 200, 3, 15
Levi McCormack, 75, 25, 4050, 300, 533
Charles McCormack, 17, -, 600, 117, 282
W. Boyd Wilson, 130, 130, 10000, 400, 600
Lewis Sulivan, 25, 75, 400, 3, 125
Hiram Casey, 110, 170, 350, 60, 378
Simon Henry, 15, 13, 100, 50, 120
George Row, 30, 20, 1500, 60, 233
Lemuel Hatten, 8, 90, 600, 3, 150
William Keyser, 40, 210, 3000, 20, 125
Wm. Hatten, 70, 30, 3000, 5, 213
Edward Burks, 50, 150, 2000, 25, 230
Joseph Payne, 75, 325, 6000, 35, 300
William Dixon, 40, 160, 2000, 20, 115
Joseph Keysor, 50, 94, 2100, 5, 98
William C. Hatten, 50, 75, 2000, 5, 105
John L. Zigler, 75, 75, 2500, 10, 276
Daniel Barnhart, 50, 40, 1200, 10, 100
John S. Hutchison, 153, 845, 5000, 50, 615
Solomon Irons, 25, 79, 1000, 5, 207

Wesly Harman, 30, 290, 1000, 3, 180
Lewis Dilman, 20, 45, 300, 3, 52
Levi Hatten, 25, 28, 500, 10, 109
Thomas Keysor, 75, 173, 3000, 50, 301
John Jones, 35, 375, 2000, 5, 150
Calvary Fulerton, 4, 5, 200, -, 150
James Wilex (Wilcox), 70, 130, 2000, 17, 300
Jesse Toney, 100, 175, 3000, 20, 715
John Toney, 125, 75, 1000, 5, 260
Able Segar, 15, 100, 450, 75, 246
Frederick A. Haddon, 50, 20, 1000, 150, 140
James Perdue, 65, 200, 3000, 15, 265
James T. McKeana, 200, 475, 4750, 100, 530
A. B. McKeana, 50, 30, 2500, -, 624
Nelson G. McCanell, 75, 125, 3300, 40, 577
John Bravo, 100, 175, 3000, 3000, 891
Thomas L. Jordan, 70, 131, 4500, 50, 500
Wm. Isaacs, 60, 375, 3000, 50, 445
Payton Staley, 60, 235, 1500, 30, 348
Wm. Swanson, 22, 58, 400, 9, 33
Charles Thacker, 40, 60, 1000, 4, 50
Zachariah Rigg, 60, 140, 1000, 15, 372
Solomon T. Staley, 100, 600, 2200, 15, 429
Granville D. Shingleton, 20, 61, 480, 50, 199
Jacob Staley, 40, 40, 2000, 15, 407
Newton Christian, 30, 120, 700, 4, 430
John W. Deskins, 35, 70, 400, 40, 142
Jeremiah Ferguson, 15, 17, 400, 5, 144
Harrison Smith, 95, 380, 1500, 80, 325

Thomas Hutchison, 35, 117, 1500, 7, 283

Adolph Osner, 38, 845, 2300, 100, 330

Absolom Balenger, 200, 800, 4000, 120, 734

Wm. J. Smith, 200, 350, 3000, 300, 778

Andrew Christian, 30, 20, 250, 3, 107

Alexander Murph, 46, 180, 1000, 10, 169

Peter Hazlett, 75, 200, 1500, 10, 193

Marshal Cyrus, 40, 60, 1000, 5, 295

Elias Hensley, 40, 130, 500, 3, 220

Ecalatis Johnson, 60, 103, 1000, 15, 277

Harrison Thacker, 180, 900, 4200, 50, 785

Harrison Ward, 45, 350, 1000, 5, 180

Edmond Hatton, 50, 75, 1000, 10, 135

Philip Hatton, 60, 150, 1000, 10, 148

Alen Newman, 200, 700, 3500, 80, 1175

John Ferguson, 75, 75, 1200, 10, 485

John D. Gilkison, 46, 269, 2500, 15, 540

George Johnson, 25, 55, 600, 4, 242

Smith Cyrus, 100, 200, 1500, 75, 405

Eligah Hatton, 40, 260, 750, 35, 397

Lane Shannon, 24, 100, 1500, 5, 272

Montague Newman, 50, 250, 2000, 20, 434

Abraham Cyrus, 150, 20, 10000, 325, 12512

Roswell Cyrus, 157, 75, 8000, 248, 2053

William Cyrus, 100, 125, 3000, 20, 724

Henry Born, 40, 120, 2600, 150, 230

Josh Chadwick, 37, 57, 4850, 50, 140

John Chadwick, 47, 153, 1800, 75, 197

Wayne Plymale, 80, 200, 2000, 75, 516

__cy Hatton, 150, 3654, 9725, 30, 632

Benj. F. Wilson, 125, 275, 3500, 60, 711

Richard B. Brown, 20, 80, 1000, 20, 183

John Plymale, 200, 570, 1000, 200, 973

J. B. Malcum, 120, 250, 3500, 100, 428

Anthony Plymale, 120, 650, 6000, 150, 1297

Samuel Kilgore, 100, 300, 3000, 110, 420

John B. Bowen, 100, 246, 6000, 50, 565

Thomas Adams, 100, 200, 6000, 20, 686

Wm. Owens, 26, 80, 500, 10, 178

John Hoback, 75, 141, 1600, 63, 536

Wm. R. Ward, 18, 72, 600, 5, 87

Henry Barber, 25, 105, 600, 10, 85

Samuel Wellman, 200, 3000, 1500, 15, 550

A. C. Handly, 43, 123, 4000, 80, 505

William Isaacs, 60, 240, 1650, 50, 810

John Adams, 70, 180, 1500, 75, 258

John Harman, 50, 50, 1500, 61, 246

Joseph Staley, 21, 40, 800, 10, 208

E. R. Staley, 25, 105, 295, 5, 70

Garret P. Rigg, 100, 200, 1500, 40, 375

Charles Hacker, 22, 155, 585, 5, 190

Strother Hatton, 30, 10, 600, 27, 2200

Benj. Davis, 60, 300, 2000, 25, 531

John Strahan, 25, 25, 300, 3, 167

Samuel Hatton, 126, 489, 5000, 40, 1528

Dehila Gilkison, 55, 12, 1000, 12, 169

Hatton A. Chaffin, 30, 40, 700, 5, 400

John Lott, 10, 153, 400, 3, 79
Goodwin Lycons, 16, 154, 1000, 4, 140
James Crockett, 25, 75, 500, 1, 103
David Peery, 25, 35, 350, 5, 77
Samuel Bilups, 80, 232, 3000, 20, 384
Levi Ballaney, 20, 580, 1500, 5, 168
James S. Parks, 50, 600, 3000, 10, 139
Wm. Walker, 35, 410, 500, 5, 550
Nathan Roberts, 18, 82, 500, 5, 100
James Walker, 35, 45, 300, 5, 200
Allen Wellman, 30, 88, 1000, 5, 167
John Grizzle, 40, 88, 500, 10, 148
Edmond Ferguson, 100, 150, 650, 5, 107
Thomas M. Ferguson, 25, 25, 300, 8, 240
Solomon Crabtree, 50, 210, 1200, 10, 210
Hiram Crabtree, 30, 158, 800, 5, 131
Garred See, 50, 50, 650, 5, 20
E. Akers, 60, 515, 1300, 5, 113
Wm. Bartrum, 120, 380, 7000, 50, 175
James Wellman, 30, 50, 1500, 75, 335
Ezekiel Roberts, 65, 135, 2000, 60,175
James Wellman, 30, 30, 1300, 50, 118
Elisha Wellman, 30, 50, 1000, 25, 310
Alex. Artrip, 15, 25, 200, 5, 115
Haines Artrip, 20, 125, 1000, 5, 86
David Webb, 30, 20, 1000, 10, 140
Alexander Wilson, 75, 140, 2000, 35, 555
Kast. Frasher, 30, 70, 700, 2, 100
William K. Frasher, 22, 30, 400, 3, 260
Harry Frasher, 25, 30, 500, 3, 152
Flemore Thompson, 80, 120, 1000, 40, 246

John Robertson, 45, 280, 1600, 25, 343
Warren Robertson, 75, 150, 200, 65, 247
Elizabeth Pauley, 19, 70, 450, 5, 170
Granvill Thompson, 45, 75, 750, 5, 245
Enoch Cyrey (Cyrus), 50, 132, 600, 6, 180
Leander Wilson, 30, 145, 400, 4, 194
Wm. Jarrel, 25, 245, 500, 15, 241
Elijah Lambert, 14, 26, 400, 12, 85
Isaac Lambert, 35, 100, 400, 4, 157
George Piles, 50, 175, 1000, 25, 301
Ruben Adkins, 30, 50, 500, 5, 200
Henry Adkins, 50, 275, 1000, 105, 227
George Adkins, 100, 600, 2500, 250, 1064
John Topen, 30, 100, 1000, 6, 85
Alex. Spence, 20, 105, 265, 10, 93
John Queen, 35, 265, 750, 10, 255
Absolem Queen, 40, 160, 1000, 5, 15
Daniel Witcher, 45, 205, 800, 10, 150
James Queen, 40, 310, 900, 4, 92
Henry Queen, 30, 220, 500, 5, 252
Lewis Mainard, 70, 50, 500, 5, 235
Wm. Ramy, 25, 40, 300, 3, 181
Thomas Wiley, 22, 78, 300, 5, 11
George Pack, 50, 450, 90, 100, 336
James Pack, 15, 100, 300, 3, 221
John Eliott, 20, 100, 400, 3, 110
Thomas McCoy, 25, 300, 700, 5, 267
John Dyer, 40, 85, 500, 10, 162
James Dunahoo, 60, 150, 1000, 10, 155
Hugh Baily, 40, 85, 200, 10, 209
Adonis Carter, 100, 375, 4500, 110, 450
Wm. Toney, 30, 100, 700, 10, 225
Joshua Cyrus, 200, 1300, 1200, 200, 876
Samuel Clark, 35, 125, 500, 10, 240
Caleb Clay, 60, 280, 1500, 5, 206
Elisha Kenderic, 30, 170, 500, 5, 127

Arthur Hobbs, 35, 175, 600, 30, 238

Peter Hoozier, 18, 94, 650, 5, 141

Peter Loaz (Loar), 20, 300, 1500, 195, 1400

Asa Booton, 60, 840, 2000, 125, 1400

Elias Childers, 50, 430, 1600, 50, 540

Milington Adkins, 100, 688, 2500, 112, 415

Hugh Ross, 50, 340, 1000, 12, 275

John H. Watts, 80, 400, 1200, 30, 433

Jacob Piles, 35, 215, 1000, 10, 243

Harvy Ferguson, 30, 40, 200, 10, 289

David Holt, 20, 25, 450, 5, 147

Wade Lambert, 35, 75, 450, 6, 125

John Casady, 8, 166, 800, 4, 50

Wesley Crabtree, 40, 390, 1200, 5, 166

Wesley Ball, 25, 260, 500, 5, 122

Lewis Bartrum, 45, 742, 1000, 5, 429

Abraham Owens, 25, 167, 500, 5, 179

Tolbert Huff, 15, 35, 200, 3, 151

Thomas Osburn, 60, 300, 1000, 60, 213

John Osburn, 30, 150, 500, 4, 127

Alexander Christian, 60, 215, 1000, 75, 318

Leander Osburn, 45, 355, 800, 40, 309

Joseph Dean, 50, 269, 1200, 35, 351

A. M. C. Davis, 85, 115, 1600, 187, 737

Samuel Ferguson, 26, 74, 400, 5, 73

Samuel Osburn, 18, 102, 400, 10, 227

Cinthia Ferguson, 40, 108, 900, 13, 329

John B. Stephens, 100, 170, 2000, 95, 497

James Dickerson, 12, 75, 3000, 8, 125

Allen W. Jackson, 12, 74, 3000, 5, 35

Samuel Osburn, 25, 75, 1000, 4, 107

Covington Ross, 45, 385, 2000, 10, 352

Jane Watts, 25, 75, 400, 4, 507

Harry Adkins, 20, 40, 400, 10, 150

Henry Ross, 10, 80, 200, 5, 81

Aaron Asbery, 30, 90, 400, 50, 190

Wm. Asbery, 50, 110, 600, 5, 109

Atison Adkins, 150, 75, 1000, 5, 321

Lewis Ferguson, 100, 1450, 3577, 225, 444

Harrison Walker, 100, 20, 400, 75, 393

James Ferguson, 40, 300, 10000, 100, 1122

Simpson Ferguson, 85, 360, 1000, 5, 92

Lemack Adkins, 30, 361, 2000, 75, 705

Robert Napier, 48,700, 300, 50, 220

Edmond Osburn, 18, 250, 1000, 25, 328

Anthony Ferguson, 17, 362, 500, 4, 122

David Sellards, 100, 94, 500, 5, 101

Dicy Adkins, 40, 1700, 5000, 25, 270

Youngha(Youngher) Napier, 20, 400, 200, 40, 415

Hezekiah Wiley, 40, 80, 500, 25, 300

John H. Queen, 30, 200, 1000, 25, 300

Kerick Queen, 30, 200, 600, 10, 333

William Walker, 15, 35, 200, 10, 144

John Johnson, 25, 100, 500, 5, 108

Arnold Perry, 35, 205, 600, 8, 367

James Mainard, 12, 63, 300, 5, 209

Samuel Jarrel, 13, 352, 600, 12, 76

Simon Mainard, 15, 60, 300, 10, 33

Loister Mathews, 60, 238, 800, 12, 279

Simpson Mainard, 25, 25, 400, 10, 157

Sampson Porter, 9, 75, 400, 3, 175

Stewart Porter, 30, 50, 400, 3, 137
Darrel Mathews, 80, 320, 1500, 10, 424
John Lycons, 20, 280, 900, 1, 179
Wm. Pinson, 40, 60, 400, 20, 196
William Watts, 15, 35, 300, 5, 151
G. P. Mays, 20, 100, 350, 5, 222
Jacob Lycons, 7, 43, 200, 2, 110
John Cox, 40, 340, 1000, 5, 532
James Sick, 13, 152, 400, 5, 52
Squire Scaggs, 20, 170, 600, 5, 223
John Osburn, 70, 520, 1500, 25, 240
Walter Napier, 10, 70, 300, 3, 40
A. T. Wooton, 45, 135, 1000, 40, 394
Eli. J. Wooton, 20, 40, 400, 5, 148
Paterick Napier, 75, 250, 1000, 50, 330
Samuel Isaac, 35, 150, 800, 3, 422
George Ball, 30, 70, 400, 5, 164
William Ball, 10, 60, 75, 12, 148
Allen Wilson, 150, 950, 3000, 75, 705
Henderson Huff, 40, 315, 650, 15, 181
Lorenz D. Hatton, 40, 65, 800, 5, 398
Waterson Workman, 60, 48, 1000, 70, 160
Stephen Dean, 35, 45, 500, 5, 24
Harry Workman, 19, 237, 800, 8, 190
Casander Workman, 50, 175, 1000, 5, 175
Simpson Booton, 55, 255, 2500, 35, 328
Allen Workman, 35, 265, 2000, 15, 311
Reuben Booton, 70, 80, 4000, 200, 555
John E. Ferguson, 20, 75, 440, 5, 133
Joseph Workman, 75, 325, 1800, 50, 512
Wm. M. Dean, 50, 100, 1000, 5, 115
Ezekiel Bloss, 55, 140, 3000, 120, 550

James M. Ross, 75, 100, 5000, 100, 683
Jemima Sellards, 30, 10, 400, 10, 267
John H. Coal, 8, 92, 400, 7, 71
John Jarrel, 10, 90, 400, 10, 52
Jesse Fry, 50, 50, 900, 15, 199
D. F. Watts, 40, 355, 1000, 10, 97
A. J. Firy, 9, 160, 400, 15, 125
Hezekiah Firy, 20, 200, 2000, 5, 121
Hezekiah Adkins, 50, 200, 2000, 5, 195
Owen Adkins, 45, 300, 2500,125, 409
John South, 40, 440, 1500, 3, 252
Jesse Mainard, 30, 140, 800, 3, 80
Hezekiah Finley, 40, 90, 700, 5, 590
James Finley, 15, 80, 300, 5, 70
Patrick Porter, 60, 270, 1300, 20, 455
Larkin Mainard, 23, 380, 400, 3, 173
A__ Mainard, 35, 210, 800, 5, 260
Gilbert Moore, 30, 40, 400, 2, 115
Sully Smith, 16, 100, 500, 5, 153
Samuel Damron, 20, 70, 500, 5, 136
John Neace, 20, 10, 400, 5, 199
Henry Workman, 75, 500, 1000, 3, 361
Alton Pack, 40, 150, 500, 5, 390
James Remy, 50, 300, 600, 3, 510
Wm. T. Coldwell, 40, 60, 600, 6, 425
Harrison Pack, 20, 100, 400, 5, 163
Samuel Nelson, 15, 60, 200, 5, 55
Israel Nelson, 60, 100, 800, 10, 484
Samuel Pack, 20, 120, 700, 10, 363
Isaac Nelson, 30, 390, 400, 3, 224
Wm. Nelson, 20, 60, 200, 3, 126
Meredith Burchart, 25, 235, 1000, 10, 288
Hary G. Dyre, 15, 100, 400, 5, 162
Wm. Wiley, 40, 260, 100, 10, 150
George B. Hinkle, 50, 850, 800, 5, 150
John W. Wampler, 50, 296, 800, 5, 194
Samuel Davis, 25, 55, 500, 10, 24

Wm. Davis, 30, 73, 350, 5, 78
Hugh C. Adkins, 12, 175, 300, 3, 54
James Copley, 40, 360, 1000, 10, 784
Christopher Chafin, 10, 75, 300, 3, 120
Nathan Chafin, 25, 50, 600, 3, 157
Stanly Chafin, 50, 100, 25, 50, 534
Wm. Brewten, 60, 300, 800, 10, 275
Wiley D. Copley, 75, 250, 700, 5, 610
Wm. Crum, 75, 250,700, 5, 610
Jackson Spalding, 70, 600, 2500, 25, 480
Thomas Spalding, 50, 150, 1200, 10, 360
Jesse Parsly, 70, 1700, 3000, 30, 946
Alex. Gaws, 20, -, 300, 10, 104
Joshua Marcum, 50, 800, 1300, 10, 450
Garrett Low, 45, 500, 2000, 5, 235
John W. Perry, 12, 288, 300, 7, 61
Wm. Brown, 15, 225, 500, 4, 86
Wm. Perry, 20, 200, 500, 5, 113
Joel Crum, 30, 200, 400, 3, 24
Thomas Crum, 12, 28, 250, 3, 59
Adam Crum, 35, 100, 600, 5, 212
James Romans, 75, 275, 1500, 15, 350
George Mainard, 35, 405, 1200, 10, 121
Lazarus Damron, 30, 300, 1000, 5, 69
Henry Smith, 40, 780, 2500, 5, 227
Car Noe, 35, 40, 600, 5, 181
R__ Preston, 40, 300, 1500, 25, 578
Denis Preston, 25, 100, 800, 5, 333
Henry Preston, 23, 200, 800, 6, 240
Andrew May, 40, 63, 350, 6, 41
John Ferguson, 37, 728, 910, 20, 785
Samuel Ferguson, 40, 100, 1000, 30, 494
Alexander Porter, 40, 335, 1000, 6, 217
Wm. Noe, 40, 300, 1000, 5, 174
John T. Noe, 25, 100, 500, 4, 27

Anderson Wilson, 60, 540, 2000, 40, 108
Washington Jackson, 35, 100, 400, 5, 62
Martha Straton, 40, 150, 1000, 5, 165
Harman Wilson, 12, 21, 150, 2, 98
Finley Thompson, 85, 500, 2400, 50, 462
Mark Sunner, 20, 150, 400, 30, 145
James Mathews, 18, 152, 500, 10, 167
James Lakin, 45, 105, 1200, 10, 357
Calvin Fuller, 26, 234, 1500, 11, 334
Harrison Parks, 60, 100, 1000, 15, 175
Mathew Belamy, 75, 175, 200, 3, 95
John Belamy, 20, 300, 250, 3, 91
Wm. Peery, 75, 200, 3000, 60, 143
Jacob Dean, 80, 275, 3000, 50, 434
Alvin Hatton, 75, 115, 4000, 60, 371
Linsey Smith, 200, 345, 12000, 2225, 1026
Wm. S. Belamy, 40, 60, 1200, 15, 190
Wm. Strother, 9, 30, 500, 10, 330
Martin Coffman, 70, 330, 4000, 15, 215
Fred Moore, 60, 850, 8000, 60, 392
John Brumly, 30, 4872, 1250, 45, 466
John Wellman, 40, 300, 1000, 10, 220
John L. Frasher, 75, 25, 3000, 40, 611
Robert Wellman, 100, 200, 2000, 75, 802
Lewis Frasher, 35, 30, 1000, 10, 109
Samuel Webb, 75, 200, 1000, 10, 274
Nathan Holt, 45, 1220, 4000, 60, 614
David Wellman, 75, 400, 3000, 35, 450
John Holt, 40, 200, 500, 30, 340
John Wellman, 20, 33, 1000, 5, 318
David Wilson, 25, 150, 200, 10, 618

Graham Wilson, 55, 50, 900, 10, 103

Elias Watts, 60, 90, 1000, 10, 169

Wm. Brumly, 100, 50, 1500, 150, 783

Thompson Ratcliff, 500, 25, 1000, 5, 472

John Jarrel, 60, 250, 1500, 25, 582

Stephen Thompson, 50, 250, 500, 5, 361

Layfayette Vinson, 50, 250, 1500, 10, 327

Francis Vinson, 700, 3200, 1000, 6, 54

Samuel Vinson, 50, 300, 1500, 75, 166

Edward Baisden, 50, 380, 2000, 20, 479

Wm. Ratcliff, 200, 1000, 4000, 50, 1319

Geo. Pack, 14, 86, 500, 5, 55

Ira Copley, 18, 150, 500, 4, 314

James Copley, 15, 165, 300, 3, 119

Hiram Loar, 260, 250, 15000, 150, 1575

James Rowe, 50, 350, 1500, 45, 184

Henry Hampton, 45, 219, 1200, 10, 117

John Pratt, 100, 400, 2000, 3, 403

Moses Damron, 15, 55, 3000, 4, 128

Burwell Watts, 20, 80, 400, 10, 115

Wade H. Thompson, 40, 120, 900, 5, 22

Wm. Porter, 35, 102, 600, 50, 228

Samuel Porter, 35, 365, 400, 5, 15

Stephen Marcum, 50, 350, 1000, 5, 100

George Damron, 15, 200, 425, 3, 185

Wiett E. Adkins, 30, 140, 600, 15, 335

Lazarus G. Damron, 15, 235, 600, 5, 136

Thomas Damron, 30, 195, 1000, 3, 160

William Damron, 14, 95, 300, 5, 120

Andrew Pack, 30, 220, 700, 20, 110

Andrew Moore, 10, 45, 300, 5, 81

John W. Whitt, 45, 115, 1700, 30, 509

Randle Samons, 25, 400, 400, 15, 248

Anthony Copley, 60, 640, 2000, 5, 398

Thomas Copley, 100, 350, 7000, 5, 696

Jesse Crum, 25, 15, 400, 20, 319

Wm. Crum, 25, 10, 500, 5, 379

James Step, 80, 420, 4000, 5, 100

Sylvester Marcum, 40, 1100, 3000, 15, 260

Jacob Marcum, 20, 250, 1000, 3, 232

Wm. R. Spalding, 40, 550, 1500, 20, 192

H. W. Price, 12, 75, 300, 2, 65

Van B. Marcum, 35, 1722, 3000, 10, 256

James H. Marcum, 20, 1055, 1075, 2, 97

Kelley Ferguson, 70, 600, 1500, 10, 895

Wm. Williamson, 10, 1500, 150, 2, 64

Elias Williamson, 15, 385, 400, 10, 167

Solomon Mead, 40, 160, 600, 12, 160

Nathan Perry, 23, 180, 800, 10, 227

Olden Williamson, 18, 100, 300, 10, 160

Samuel Damron, 40, 2800, 377, 50, 581

Moses D. Damron, 40, 160, 700, 7, 60

Wm. Nixon, 60, 490, 1500, 100, 512

Webster County, West Virginia
1860 Agricultural Census

The University of North Carolina at Chapel Hill filmed the 1860 agricultural census for Webster County from originals at the West Virginia State Archives under a grant from the National Science Foundation in 1963.

Columns 1, 2, 3, 4, 5, and 13 represent the following information on the census:
1. Name of Owner, Agent or Manager of Farm
2. Acres of Improved Land
3. Acres of Unimproved Land
4. Cash Value of the Farm
5. Value of Farming Implements and Machinery
13. Value of Livestock

Robert E. Given, 12, 351, 800, 1, 48
Adam Given, 12, 150, 1200, 3, 147
William Given, 100, 115, 1000, 40, 195
David Hamrick, 20, 30, 300, 15, 144
William Hamrick, 14, 174, 300, -, 88
Washington Cutless (Cutlip), 6, 113, 50, -, 40
James Hamrick, 20, 487, 1500, -, 480
John Hamrick, 7, 293, 268, -, 157
Peter Hamrick, 15, 275, 500, 10, 470
Joseph Greene, 15, 87, 300, -, 125
Daniel Lough, 3, 117, 150, -, 60
Wm. F. M. Chapman, 5, 145, 150, 5, 146
Wm. F. Strong, 10, 4000, 1000, 10, 184
Abner Cogar, 3, 150, 150, -, 3
Wm. G. Gregory, 20, 190, 1000, 5, 323
Benjamin Hamack, 5, 295, 250, 2, 100
George Dodrill, 54, 956, 1500, 15, 105
Fidelia S. Dodrill, 30, 31, 600, 10, 245
James M. Hamrick, 35, 661, 1000, 20, 500
Bernard Mollahan, -, -, -, -, 60

Robert Dodrill, 25, 264, 700, 10, 429
Christopher M. Hamrick, 75, 812, 2000, 25, 608
Widow Nancy Hamrick, 60, 624, 1500, 10, 287
Adam G. Hamrick, 75, 125, 1000, 12, 212
Currance Gregory, 40, 472, 1000, 15, 427
Adam G. Gregory, 40, 560, 1000, 20, 500
Wm. W. Tracey, -, -, -, 1, 91
Arthur McBickle, 20, 80, 400, 10, 13
Addison McHamrick, 10, 446, 500, 10, 170
Samuel F. Miller, 50, 745, 991, 5, 200
Benjamin Hamrick, 6, 105, 111, 10, 130
Patrick Care (Case, Can), -, -, -, -, 88
Charles Dodrill, 30, 229, 800, 15, 320
James M. Cogar, 5, 214, 219, 4, 100
Isaac Hamrick, 25, 75, 400, 25, 319
Samuel Griffin, -, -, -, -, 60
Isaac G. Lynch, 10, 90, 250, 8, 82
Thomas Cogar, 3, 3, 50, -, 113
Peter L. Cogar, 35, 465, 1000, 6, 236
Samuel McAvoy, 20, 280, 1000, 20, 60

Sampson Sawyers, 25, 375, 1300, 20, 487

William Given, 20, 480, 1000, 5, -

James H. Raner, 8, 186, 400, 5, 61

Robert Given, 40, 137, 800, 30, 721

Caleb A. Gardner, 15, 62, 600, 2, 124

Jackson Cutlip, 12, 90, 300, 5, 51

Samuel H. Woods, 25, 46, 500, 15, 100

William Procter, 18, 85, 800, 8, 150

Adam Rader, 60, 172, 1500, 70, 230

Levi C. Hedges (Hedger), 2, 38, 100, -, 30

John Dodds, 20, 80, 800, 10, 110

Thomas J. Morton, 35, 265, 1500, 15, 591

Robert Gregory, 18, 482, 1000, 5, 108

George W. H. Miller, 100, 2150, 400, 30, 987

John J. Grigsby, 15, 85, 500, 3, 215

John C. Payne Jr., 12, 92, 400, 3, 106

Allen Hamrick, 23, 295, 1000, 5, 197

Christopher Cogar, 25, 86, 700, 5, 158

Silas Cogar, 20, 166, 500, 2, 194

Solomon Fisher, 16, 124, 500, 6, 126

Addison Fisher, 3, 67, 100, 5, 54

Edward W. Sizemore, 9, 493, 300, 5, 78

Addison Robinett, 7, -, -, -, 12

Andrew Fisher, -, -, -, -, 21

Zechariah R. Howell, 15, 35, 500, -, 91

James W. Clifton, 30, 450, 350, 15, 194

Daniel L. Perdew, 15, 142, 600, 4, 138

James Gadd, 5, 5, 200, -, -

Edward B. Sizemore, 12, 138, 500, 5, 100

John Cool, 15, 185, 600, 3, 110

J. H. Bledsoe, 18, 982, 1500, -, -

William M. Rader, -, -, -, -, 40

Jacob P. Conrad, 80, 2920, 5000, 20, 440

Solomon Arbogast, 8, 92, 300, 2, 131

Andrew C. Buckhannon, -, -, -, -, 15

Alfred D. Anderson, 16, 311, 200, 3, 40

George D. Anderson, 14, 486, 500, 3, 110

Alexander C. Anderson, 5, 495, 500, 3, 64

Isaac Bender, 40, 160, 500, 5, 492

Thomas Bender, 10, 90, 500, 3, 132

John Bender, 6, 255, 500, -, 39

Robert McCray, 60, 2089, 3000, 20, 59

Wesley Boggs, -, 100, 100, 5, 73

Lorenzo L. W. Pugh, 5, 195, 300, 2, 58

James McCray, 20, 380, 800, 2, 174

Daniel Snyder, 20, 130, 300, 3, 40

Amos Tharp, -, -, -, -, 63

Elias Snyder, 14, 361, 750, -, 88

John A. Mace, 130, 760, 5000, 100, 574

Josiah Cowger, 30, 270, 800, 4, 270

William Cogar Jr., 12, 88, 300, 3, 110

George W. Vance, 9, 241, 400, 1, 180

A. M. Anderson, 35, 1265, 1500, 6, 236

George D. McCartney, -, -, -, -, 64

Felix Albert, 25, 175, 300, 4, 109

Junis Cogar, 25, 525, 150, 6, 193

John W. Buckhannon, -, -, -, 2, 63

Abraham Buckhannon, 106, 795, 1900, 15, 180

Jesse Cowger, 30, 698, 1000, 10, 178

John Motes, 3, 373, 281, 1, 32

Caleb Henkle, 11, 340, 800, 30, 59

John Jordon, 25, 225, 750, 3, 286

Amoshaddi Jordon, 26, 474, 1000, 15, 99

Andrew Jordon, 100, 1300, 3500, 20, 264

Addison Cutlip, 60, 314, 2000, 20, 504

Joseph Brown, 12, 185, 500, 4, 164

Lewis Cutlip, 30, 532, 600, 10, 230

William H. Mollahan, 33, 1872, 1520, 5, 262

James Mollohan, 100, 927, 1000, 10, 159

Jeremiah Brown, -, -, -, 2, 24

Edward W. Ware, 60, 106, 600, 12, 188

Leonard J. Brake, 40, 430, 1000, 10, 64

Cornelius G. Cool, 5, 185, 300, -, 116

John O. Cool, -, -, -, -, 236

Samuel Cutlip, 30, 66, 500, 10, 109

Jeremiah B. Howell, 22, 78, 300, 5, 36

John C. Perrine, 15, 585, 500, -, 102

Asa W. Fisher, -, -, -, 4, 200

John Bickle, 40, 390, 1000, 16, 304

George W. Bickle, -, -, -, -, 140

Christian B. Ware, 1, 499, 200, -, 134

Benjamin Cogar, 25, 275, 750, 5, 152

Christopher Strader, 30, 320, 700, 5, 146

William R. Arters, 25, 475, 700, 5, 133

John Fares, 50, 1000, 1500, 5, 296

Waltar Cool, 300, 1756, 12300, 75, 1152

Jonas C. Strader, 25, 255, 1000, 8, 147

William H. Cogar, 20, 735, 1000, 2, 142

Arther Hickman, 100, 400, 5000, 180, 464

William H. Woods, 14, 184, 500, 5, 62

George W. Woods, -, -, -, -, 150

John Chapman, 40, 95, 800, 10, 65

James Williams, -, -, -, 2, 83

Andrew E. Barnett, 1, 649, 650, 4, 41

Anderson Cutlip, 7, 93, 100, 2, 33

William C. Murphey, -, -, -, 8, 52

John C. Fowler, 15, 445, 460, 50, 227

Jesse Davis, 40, 60, 200, 12, 86

Jesse Davis, 40, 60, 200, 12, 86

Henry C. Barnett, 8, 117, 250, 5, 215

William C. Barnett, 25, 195,600, 5, 148

John Baughman, 50, 2802, 2852, 50, 575

George F. Lewis, -, -, -, -, 10

John L. Carpenter, 60, 1766, 2250, 20, 127

William H. Perrine, 1, 327, 30, 3, 31

Addison Carpenter, 8, 42, 1000, 3, 102

John S. Carpenter, 16, 34, 800, 3, 80

Jehu Cogar, 13, 37, 150, 4, 139

Jacob Hosey, -, -, -, -, 14

Simon Weese, 14, 76, 90, 5, 55

Sarah Greene, 15, 65, 75, -, 43

Rud (Reece) Clifton, -, -, -, -, 117

John Brooks, 11, 186, 200, 3, 133

Jesse Clifton, 20, 275, 650, -, 188

William W. Clifton, 40, 251, 1000, 10, 200

John B. McCourt, 50, 2150, 3000, 10, 623

Andrew J. G. Burnes, -, 150, 100, -, 90

Samuel C. Miller, -, -, -, -, 35

James Woods, -, 17, 85, 2, 55

James J. McCourt, -, -, -, 2, 127

Samuel C. Tharp, 10, 120, 250, 2, 35

Zechariah Woods, 40, 60, 500, -, -

Andrew Woods, -, -, -, -, 130

Lewis Tharp, 15, 110, 500, 2, 77

Robert Miller, 8, 42, 250, 1, 25

Andrew McCourt, -, -, -, 7, 39

James Pritt, 50, 430, 800, -, 277

William C. Cochran, -, -, -, 6, 49

Ezra Clifton Sr., 40, 490, 500, 4, 216

Francis Pritt, -, -, -, 4, 20

John W. Arters, 12, 311, 300, 3, 67
Adam G. Lynch, 25, 390, 600, 5, 277
William McAvoy, -, -, -, -, 25
Wilson Arther, 40, 95, 800, 10, 324
Thomas J. Miller, 60, 100, 800, 15, 439
Widow Nancy Miller, 20, 104, 700, 3, 182
Jane Dye, 45, 988, 6000, 140, 662
Jesse W. Payne, 12, 127 400, 3, 103
Archibald Cogar, 30, 291, 1000, 5, 282
Elijah Skidmore, -, -, -, -, 30
Duncan McLaughlin, 65, 2936, 1000, 150, 300
John E. Hall, -, -, -, -, 198
George Cogar, 25, 35, 500, -, 164
John Lynch Sr., 60, 20, 1000, 10, 250
John Lynch Jr., 12, 88, 600, 5, 86
John R. Cogar, 25, 143, 800, 5, 148
David Baughman, 20, 98, 500, 5, 256
William T. Hamrick, 30, 1670, 2000, 5, 525
John Miller, 100, 511, 5500, 25, 857
Addison McL. Miller, -, -, -, -, 285
Matthew Given, 30, 218, 1000, 5, 159
Francis Gardner, 25, 75, 700, 25, 123
Cutlip Myers, 9, 391, 700, 3, 183
James McAvoy, 10, 140, 500, 5, 203
James A. Adkison, 10, 130, 400, 3, 105
William C. Young, 15, 285, 600, 7, 50
Elijah D. Greene, 15, 165, 600, 2, 100
Isaac Greene Jr., -, -, -, -, 22
Isaac Greene Sr., 30, 490, 1000, 5, 55
Isaac B. Tyler, -, -, -, -, 55
Hezekiah Holcomb, -, -, -, 5, 179
Isaac Weese, 40, 60, 800, 15, 227
Abraham Goff, 20, 244, 800, 5, 134
Enos Weese, 54, 1176, 2460, 80, 595

Lewis McElwane, 30, 330, 1000, 5, 189
William C. Dodrill, 30, 173, 1250, 20, 379
Samuel Holcomb, -, -, -, 5, 36
Benoni Griffin, 40, 1060, 2000, 70, 325
Joseph M. Perrine, 40, 610, 2000, 12, 305
James M. Barnett, -, -, -, -, 55
Jacob Carpenter, -, 675, 675, 4, 156
Isaac J. Sawyers, 12, 138, 700, 5, 86
Benijah Freeman, -, -, -, 7, 186
Andrew McElwane, 10, 410, 1000, 5, 88
Thomas M. Reynolds, 30, 270, 1000, 50, 315
Robert L. Henderson, -, -, -, -, 167
John J. Tracey, -, -, -, -, 20
Ezra B. Clifton, 7, 168, 275, 5, 37
George M. Sawyers, -, 200, 200, 10, 36
John McGwire, 7, 68, 200, -, 98
Shedrach C. Woods, 25, 113, 400, 10, 155
John L. Morton, 30, 200, 1000, 100, 359
Adam Gregory, 7, 153, 300, 10, 122
Beri McCourt, -, -, -, 5, 25
James R.Gum, 35, 374, 1000, 5, 141
Henry C. Moore, 32, 22000, 8000, 25, 468
John Townsend, 40, 260, 1000, 100, 260
John C. Payne Sr., 3, 17, 50, 15, 25
John G. Given, 40, 235, 1500, 20, 414
Margaret E. Hollister, 50, 5383, 2500, 100, 264
William F. Gardner, 30, 41, 500, 8, 81
Edward Morton, 18, 382, 1500, 10, 396
George W. Morton, 35, 365, 1500, 10, 277
James Hanna, 30, 170, 450, 5, 168

Samuel Given, 150, 13207, 7500, 80, 360

Septimius M. Board, 50, 2250, 4000, 10, 970

Wetzel County, West Virginia
1860 Agricultural Census

The University of North Carolina at Chapel Hill filmed the 1860 agricultural census for Wetzel County from originals at the West Virginia State Archives under a grant from the National Science Foundation in 1963.

Columns 1, 2, 3, 4, 5, and 13 represent the following information on the census:
1. Name of Owner, Agent or Manager of Farm
2. Acres of Improved Land
3. Acres of Unimproved Land
4. Cash Value of the Farm
5. Value of Farming Implements and Machinery
13. Value of Livestock

This county is very faint throughout both names and numbers.

_____ Layhoff, 15, 2050, 4140, 6, 12
_____ Petit, 42, 55, 810, 75, 207
Thompson Ulm, 60, 231, 1416, 20, 65
_____ Marshal, 100, 500, 10000, 50, 115
P____ Sole, 150, 430, 6100, 150, 669
_____ Marshall, 40, 60, 600, 10, 331
John Cremeans, 80, 108, 610, 4, 75
_. _. Sole, 7, 47, 360, 10, 115
James Woods, 33, 187, 1540, 18, 225
Hines Church, 36, 166, 1900, 25, 200
George Krichserburg (Hicksenbaugh), 60, 140, 2000, 50, 366
Frances Church, 30, 70, 1000, 23, 211
Joseph A. Crane, 40, 127, 1450, 25, 226
Chery Church, 25, 95, 1000, 3, 195
__alk Hotestutter, 71, 100, 410, 26, 211
Adrian Villers, 25, 167, 642, 10, 75
_____ Roberts, 7, 32, 2110, 10, 95

Samuel Church, 25, 215, 1800, 10, 51
John Staltmier, 5, 45, 225, 10, 41
John Hicksenbaugh, 6, -, 100, 5, 25
_____ Sole, 35, 105, 1400, 15, 162
Saml. Staltmier, 3, 137, 186, -, 160
David Staltmier, 100, 700, 3500, 25, 433
James Staltmier, 3, 137, 1121, -, 125
_____ Statlemier, 25, 81, 1000, 15, 18
M___ Villers, 4, 100, 600, -, -
Margaret Brown, 17, 107, 1250, -, 25
Levi Hays, 100, 300, 3200, 25, 245
Daniel Remley, 20, 70, 120, 5, 20
William P. Dawson, 18, -, 150, -, 21
George Lemley (Remley), 40, 17, 855, 5, 208
John H. Sprague, 65, 280, 4000, 16, 238
John Akers, 65, 280, 4000, 16, 210
Thomas Ashby, 40, 52, 900, 20, 200
Jams Holenbrick, 35, 65, 600, 10, 80
William Morrison, 20, 23, 245, 5, 50
George W. Butcher, 25, 25, 500, 10, 125
Robert Butcher, 50, 50, 1000, 10, 150

Matthew Butcher, 30, 35, 650, 10, 92

_. _. Helmbrick, 41, 19, 710, 10, 143

John _. Hilesrutter, 301, 225, 4725, 210, 1297

_____ Helmbrick, 10, 20, 400, 2, 75

Saml. McDonald, 86, 4, 100, -, 25

_____ Helmbrick, 45, 80, 1120, 11, 100

_____ Anderson, 15, 85, 611, 10, 150

Henry Church, 20, 40, 480, 10, 200

Henry Wright, 21, 41, 480, 5, 41

Absalem Wright, 50, 230, 3800, 95, 200

Joseph R. Jeffirs, 21, -, 100, -, 21

John Lazear, 110, 115, 2000, 50, 365

Peter J. Bailrey (Bailrug), 30, 38, 1111, 2, 154

_____ Dawson, 71, 92, 2111, 36, 330

__than Glover, 64, 40, 1000, 15, 330

Thomas Hoge, 130, 71, 400, 10, 544

Stephen Roberts, 35, 91, 1250, 75, 299

James Horner, 15, 45, 200, 10, 90

Eugenius Bray (Troy), 100, 121, 41150, 51, 364

William Hasker, 55, -, 511, 50, 311

Charles Horner, 65, 150, 3500, 35, 350

Joseph Morris, 35, 65, 1000, -, 365

Saml. Barburg, 14, 13, 321, -, -

John Roberts, 7, -, 11, -, 113

Christian Bartrug, 15, 35, 25, -, 20

James Glover, 50, 165, 2159, 35, 335

_. P. Philips, 6, 45, 251, -, 31

Michael Hindgarden, 95, 95, 2150, 31, 160

John _. Hindgarden, 15, 56, 350, 11, 203

Ulrich R. Shindler, 35, 109, 1040, 25, 213

_. Cunningham, 30, 14, 1000, 20, 123

Ephraim Glover, 9, -, 91, -, 12

James Anderson, 51, 75, 1311, 21, 265

Enoch Hickenbotten, 80, 121, 1000, 15, 273

Benj. Crenshaw, 31, 156, 900, 18, 100

Joseph Thomas, 36, 70, 600, 15, 161

James Horner, 15, 35, 351, 5, 125

Jacob Faught, 20, 41, 411, 100, 191

John Byard, 5, -, 40, 2, 37

Saml. T. Garden, 70, 200, 2161, 25, 381

Joseph Park, 25, 59, 500, 15, 143

Jacob Stoneking, 40, 90, 715, 15, 300

_____ Park, 30, 70, 500, 10, 132

_____ Bell, 75, 325, 1600, 25, 150

Enoch Roberts, 50, 112, 1200, 20, 273

Thomas Floyd, 12, 43, 300, 5, 741

_. McLaughlin, 40, 65, 450, 10, 250

Elias Villers, 20, 50, 400, 10, 125

James Villers, 50, 150, 600, 15, 225

Jesse Stewart, 100, 230, 2000, 50, 46

Mary Barbrug, 50, 7150, 2500, 50, 287

Peter P. Barbrug (Bartrug), 25, 75, 1000, 10, 180

_____ Bartrug, 100, 300, 4000, 10, 150

_. H. Wright, 20, 180, 1200, 5, 65

Johnathan Shreve, 50, 350, 2000, 75, 554

Absalom Willey, 50, 200, 2000, 15, 245

_____ Shreve, 25, 100, 1000, 10, 175

Jeremiah King, 12, 8, 150, -, 24

Edmund Kyle, 83, 1595, 4000, 5, 332

John J. Kyle, 10, 214, 500, 10, 60

Stephen Downey, 50, 1970, 3000, -, 36

James G. West, 50, 3500, 12000, 31, 500

Addison West, 30, 1170, 3600, 10, 300

Danul Ash, 300, 200, 1100, 5, 200

Aaron Snodgrass, 25, 195, 1220, 5, 135

John D. Snodgrass, 40, 300, 1050, 10, 309

Benj. J. Freeland, 35, 94, 700, 10, 290

Levi Hays, 50, 194, 3000, 10, 218

Edmund J. Hays, 20, 145, 495, 5, 117

Levi Starkey, 100, 394, 4000, 125, 684

Levi Starkey Jr., 30, 57, 800, 9, 132

Aberriman Rice, 40, 150, 600, 10, 236

William Price, 30, 140, 450, 5, 156

Saml. Price, 30, 18, 864, 5, 767

John Price, 20, 75, 450, 5, 92

James Starkey, 25, 89, 300, 5, 75

William Gump (Stump), 6, 125, 500, -, 40

William Norris, 25, 167, 496, 5, 114

J. T. C. Copenhagen, 12, 45, 114, 2, 113

David Starkey, 14, 18, 300, 10, 120

G. T. Sniger, 3, 20, 200, -, 24

Zac Kinder, 25, 165, 800, 5, 30

Joseph Morgan, 50, 150, 1500, 75, 475

Pinson Cain, 4, 100, 200, -, 20

William Baker, 12, 100, 400, 51, 100

David Barker, 10, 23, 150, 5, 150

James Musgrove, 75, 2025, 40000, 100, 467

Saml. Starkey, 35, 25, 150, 5, 100

William Strait, 30, 20, 300, 10, 175

Allen Edgell, 25, 75, 1000, 51, 90

John Haddox, 7, -, 70, 5, 115

Samuel Starkey, 10, 90, 200, 5, 100

Levi Starkey, 20, 30, 250, 5, 100

Henry Talkington, 100, 122, 3000, 40, 492

David Talkington, 50, 50, 1000, 10, 129

Patrick Emanuel, 135, 165, 1000, 25, 165

David Trader, 20, 80, 300, 5, 115

Abram Ice, 30, 250, 270, 5, 50

Elias Shreve, 20, 149, 507, 5, 106

J. C. Cunningham, 18, 182, 630, 5, 60

John _. Brumel, 60, 5940, 1800, 300, 200

John Boner (Bones), 4, 135, 500, 5, 60

John Savon, 15, 115, 540, -, -

Jesse Morris, 16, 20, 100, 5, 250

Alexander Lantz, 200, 5766, 14415, 150, 1185

Carul Willey, 40, 112, 500, 10, 218

John Watson, 31, 97, 300, 10, 82

Richard Anderson, 75, 150, 2300, 200, 527

E. M. Hays, 100, 180, 2800, 100, 90

John Hays, 40, 127, 1000, 25, 383

Pery West, 100, 830, 4000, 25, 929

Zac Wates, 45, 255, 1895, 10, 249

Zacker Wates, 60, 185, 2500, 16, 182

A. B. Ice, 50, 90, 800, 40, 206

Silas Wates, 15, 135, 500, 10, 120

Jesse Ice, 30, 100, 500, 12, 190

John Keller, 25, 100, 500, 10, 125

Abram Barker, 9, 50, 300, -, 75

James Edgell, 30, 70, 300, 10, 100

John Ice, 55, 279, 1000, 5, 200

John O. King, 15, 275, 2250, 10, 140

Streeter Willey, 35, 220, 1800, 10, 70

Boston (Barton) _. Clerag, 50, 100, 1000, 12, 162

Samuel Bland, 40, 122, 800, 11, 207

J__ King, 50, 90, 1500, 15, 388

Augustus Wates, 25, 115, 520, 10, 315

William Willey, 50, 200, 1600, 10, 310

James Hickenbottom, 60, 405, 2790, 15, 331

Hammond Hickenbotham, 100, 600, 4800, 100, 396

W__- Grump, 2, 56, 180, 2, 20

Samuel Poe, 20, 45, 500, 25, 258

Joseph Higgins, 60, 1500, 3320, 5, 70

Johnathan Higgins, 8, 100, 800, 5, 70

Frederick Grump, 12, 75, 300, 5, 80

Hiram Grump, 12, 100, 400, 5, 30

Samuel Snyder, 9, 91, 500, 5, 30

James Jolliff, 45, 405, 3000, 8, 295

Hezekiah Jolliff, 50, 250, 1800, 50, 278

Benjamin Martin, 175, 150, 4000, 150, 866

Samuel Long, 140, 938, 4213, 75, 921

William Mattheny, 125, 572, 5744, 100, 650

John W. Morgan, 140, 3160, 10500, 5, 335

Barton Oliver, 40, 40, 1000, 40, 135

Elihu Morgan, 7, 1, 250, -, 175

Joseph Hunter, 15, 35, 500, 5, 120

James Savage, 20, 30, 1000, 10, 30

Stewart Hall, 25, 95, 900, 10, 200

Francis Lutz, 30, 30, 600, 10, 150

George Lutz, 30, 30, 500, 10, 147

Gabriel Strait, 30, 45, 1000, 10, 100

Enid Fleharty, 30, 170, 800, 12, 13

Joseph Cambridge, 30, 170, 800, 12, 20

George Parks, 50, 1050, 2200, 15, 180

Levi Low, 100, 600, 6000, 170, 1770

William Noland, 75, 505, 1244, 75, 669

Sabias Noland, 15, 100, 300, 20, 90

Henry Noland, 30, 70, 500, 20, 225

Samuel Weekley, 15, 45, 300, 5, 125

Wm. Fleharty, 20, 130, 300, 5, 150

Jas. Arnett, 25, 175, 600, 10, 250

Elias Stoneking, 40, 43, 600, 90, 187

John Stevens, 8, 92, 600, 10, 140

Joseph Lemasters, 4, 96, 300, 14, 20

Lassy Baker, 50, 90, 1000, 20, 225

James Lemasters, 25, 125, 800, 10, 200

Jas. Glivens, 20, 480, 1000, 15, 192

Augustus Strait, 25, 75, 550, 8, 320

L__ Booth, 25, 275, 2000, 10, 245

J___ Lazear, 20, 100, 500, 10, 75

_. H. Tanery, 40, 90, 600, 5, 250

John Keller, 25, 100, 500, -, 125

Charles Flaherty, 25, 125, 180, 5, 100

_. R. Wetzel, 10, 90, 300, 5, 50

E. G. _. Flaherty, 10, 90, 300, 5, 50

William Kiger, 19, 81, 150, 5, 103

Josiah Stram, 13, 63, 350, 5, 104

Charles Kiger, 34, 80, 800, 10, 160

William Meyers, 25, 500, 2000, 10, 150

Jeremiah King, 65, 385, 3000, 35, 348

Thomas P. King, 30, 34, 800, 50, 325

H. W. Lippenwell, 10, 21, 800, -, 60

Mary H. Roberts, 6, 100, 500, 5, 100

John Allen, 100, 231, 4000, 75, 828

Elijah Becket, 10, 100, 200, -, 100

Joseph Harris, 5, -, 125, -, 219

Elisha Morgan, 35, 800, 3200, 15, 274

C. J. Morgan, 40, 80, 1200, 80, 475

A. J. Morgan, 35, 800, 3200, -, 20

Morgan Morgan, 100, 120, 4000, 75, 400

Thomas McGoven, 25, 75, 1500, 25, 300

Wm. A. Clepstine, 75, 425, 3000, 100, 336

E. J. Rodgers, 11, -, 200, 10, 46

Jacob Flaherty, 40, 100, 2000, 9, 147

John West, 40, 260, 60, 5, 402

Johnathan Cashran (Cochran), 28, 46, 375, 5, 184

John Henderson, 12, 100, 375, 5, 975

Robert Cunningham, 40, 200, 100, 4, 285

William Flaherty, 30, 70, 700, 12, 192

Anthony Headly, 60, 310, 1400, 20, 500

John Martin, 50, 970, 4000, 75, 985

John J. Lantz, 130, 900, 8000, 200, 1000

_____ Algers, 4, 5, 50, 30, 100

Thomas K. Strait, 50, 362, 12000, 20, 307

Enoch Cunningham, 3, 175, 530, 2, 165

Elijah Stacker, 20, 80, 400, 4, 70

John Kennedy, 8, 430, 1200, 3, 95

J. H. Bassel, 15, 50, 250, 1, 100

Elisha McCormack, 25, 128, 500, 3, 143

Elias Hickman, 8, 190, 600, 3, 96

Elisha Ferrell, 16, 97, 485, 10, 150

Frederick Fox, 20, 150, 700, 10, 100

Margaret Malcel, 20, 39, 600, 10, 90

Wm. A. Williams, 30, 60, 1000, 10, 269

James A. Williams, 32, 68, 1000, 15, 85

James Cochran, 125, 675, 7500, 260, 1114

Jacob Miller, 50, 242, 3500, 10, 98

Carter Headley, 10, 214, 1125, 6, 125

Samuel Henderson, 2, 25, 54, 5, 33

David Springer, 14, -, 280, 3, 174

_. Stevens, 60, 140, 1600, 100, 629

Isaac Steel, 60, 20, 250, 10, 300

Hezekiah Ally (Allen), 115, 135, 4000, 100, 639

__ry Flaherty, 20, 30, 300, 10, 165

Eli Long, 40, 60, 1000, 10, 175

William Little, 13, 187, 3000, 20, 15

William Wade, 25, 105, 6000, 10, 110

Levi Long, 50, 200, 1500, 40, 100

James Robinson, 25, 150, 1500, 12, 188

John Showalter, 15, 116, 800, 10, 25

John Shusnider, 40, 106, 2000, 10, 120

Hezekiah Wade, 30, 17, 1500, 15, 256

Cephas McMasters, 20, 80, 500, 10, 75

Mary Yancey, 25, 30, 500, 10, 52

Levi Anderson, 130, 200, 1500, 75, 200

William Wade, 15, 87, 500, 5, 150

Peter Glover, 100, 162, 2630, 15, 436

Theophilus Horner, 40, 100, 1500, 10, 75

Thomas Stansberry, 30, 133, 1000, 15, 250

Jacob Kirkpatrick, 72, 225, 2000, 75, 622

Anthony Lemasters, 30, 67, 2000, 20, 150

_____ Anderson, 30, 130, 1600, 75, 318

Asbury Crow, 10, 7, 41, 10, 160

Nancy Stansberry, 25, 117, 800, 5, 130

_____ Gerley, 18, 100, 500, 5, 62

Enoch Lemasters, 9, 12, 120, 5, 58

Damson Wade, 8, 100, 1000, 10, 50

Solomon Slide, 18, 182, 1800, 5, 60

_____ Lemasters, 18, 100, 500, 5, 145

Nathan Little, 30, 79, 1290, 5, 120

_____ Carmey, 50, 400, 2700, 30, 337

Phoebe Morris, 15, 115, 1000, 5, 150

Thomas Bartlett, 10, 26, 400, 5, 22

Baker Postlewate, 20, 107, 400, 5, 20

R. C. Morgan, 200, 300, 5000, 100, 310

Emanuel Robinson, 120, 1155, 8000, 100, 655

Thomas Sheppard, 50, 69, 2000, 75, 410

Thomas Delaney, 25, 112, 500, 10, 178

Thomas Postlewate, 75, 315, 1650, 50, 370

Saml. Taylor, 12, 150, 700, 200, 405

John Sole, 200, 60, 3400, 50, 492
Adam Stump (Grump), 200, 60, 3400, -, 250
Crawford Moore, 50, 50, 900, 30, 175
Frank Jackson, 18, 42, 300, 5, 180
Isaac Moore, 30, 63, 600, 2, 152
John Delaney, 90, 170, 1800, 75, 717
Emanuel Postlewate, 100, 365, 3000, 70, 448
George Loudenslager, 50, 125, 1000, 25, 225
John Delaney, 50, 127, 1000, 20, 102
James Goodrich, 18, 62, 300, 10, 42
_____ Miller, 10, 80, 300, 5, 110
John Lemasters, 8, 92, 300, 5, 100
_____ Lemasters, 40, 45, 1000, 10, 215
Margaret McBride, 25, 47, 350, 5, 300
_. B. Gorbey, 45, 145, 1200, 42, 368
James Morris, 75, 224, 2500, 5, 105
Johnathan Morris, 75, 319, 2500, 5, 137
Nathan Morris, 8, 60, 400, 5, 75
Amos Morris, 50, 175, 1750, 5, 50
Johnathan Morris, 50, 269, 3000, 10, 270
Wm. McHenry, 75, 125, 1600, 50, 262
_____ Miller, 90, 163, 2000, 3, 515
_____ Rider, 36, 114, 1000, 31, 219
_____ Lemasters, 50, 90, 2000, 15, 313
J___ Camery (Carney), 125, 522, 4550, 75, 609
Solomon Carney (Camery), 15, 88, 500, -, 162
John Bland, 4, 75, 300, 5, 25
Isaac Bartlett, 30, 70, 600, 5, 41
Eli Miller, 50, 90, 1000, 15, 200
William Miller, 30, 155, 1000, 15, 295
Christian Showalter, 30, 58, 500, 15, 169
_____ Long, 40, 260, 3000, 20, 320

Joseph Hamond, 25, 75, 700, 60, 200
Isaac Miller, 75, 333, 3000, 75, 334
Peter Kirbey, 25, 95, 1000, 10, 25
John Wade, 30, 45, 500, 10, 117
Stephen Bloid (Bland), 15, 25, 250, 10, 60
Felix Piles, 20, 115, 405, 5, 150
Perry Jonson, 45, 133, 1504, 30, 291
Ashbery Showalter, 20, 75, 500, 10, 200
Joseph Jackson, 16, 145, 495, 10, 140
John H. Sterns, 10, 40, 400, 10, 60
John Cane, 40, 110, 900, 15, 176
_____ F. Clark, 7, 60, 240, 5, 105
Isaac Baker, 15, 100, 800, 3, 110
Henry Dorothy, 15, 7, 88, 10, 110
Wm. W. Sears, 10, 190, 1400, 5, 200
William Woods, 60, 100, 1200, 10, 163
Isaac Hults, 80, 65, 1600, 60, 300
_. Chambers, 13, 52, 900, 10, 150
Jno. Jolliff, 50, 100, 7000, 10, 285
George W. Davis, 30, 62, 400, 10, 170
Abraham Earlgrieve (Eastgrieve, Eastgreene), 13, 26, 200, 5, 60
William B. Allen, 20, 86, 700, 10, 180
Thomas Woods, 12, 52, 500, 10, 27
George Gain, 20, 64, 400, 5, 15
Presly Strowsnider, 20, 58, 500, 3, 100
Peter Coen (Corn), 50, 50, 1200, 29, 250
Thomas Delaney, 25, 75, 800, 20, 250
William Shuman, 100, 75, 2000, 20, 450
Jesse Shuman, 40, 80, 2000, 15, 429
John _. Delaney, 40, 90, 1000, 25, 178
James Willey, 30, 70, 800, 10, 135
Adam Loudenslager, 40, 130, 1000, 40, 367
Henry Utt, 40, 160, 1200, 5, 220

_____ Sole, 40, 230, 1000, 40, 200
Josephus Henderson, 40, 60, 1000, 75, 192
James Anderson, 75, 75, 1550, 8, 400
William Hibles, 75, 180, 1700, 25, 175
David Anderson, 40, 104, 2000, 100, 548
Thomas McMasters, 15, 35, 450, 5, 108
Enoch Anderson, 45, 65, 1000, 15, 276
Elihu Guthrie, 40, 60, 650, 14, 150
_. B. Anderson, 90, 40, 1500, 200, 407
_. Anderson, 30, 70, 1200, 20, 366
Alexander Ulmer, 30, 65, 1100, 40, 365
Nicholas Cross, 80, 100, 1600, 100, 178
William Cross, 60, 40, 1000, 18, 154
John Murphy, 35, 95, 800, 25, 208
Nathan Carney, 100, 85, 2000, 75, 500
Aaron Asher, 40, 60, 1000, 15, 209
John Anderson, 125, 175, 3000, 150, 577
George Sole, 25, 25, 450, 25, 247
George Vanhorn, 30, 53, 500, 38, 150
John Barker, 40, 82, 700, 5, 182
Eugenius Hibbs, 10, 100, 10, -, -
Charles Strosnider, 11, 62, 400, 12, 182
James Thomas, 80, 60, 2000, 15, 18
John Robinson, 30, 120, 1140, 1, 200
Thomas Read, 4, 51, 275, 2, 20
Sarah Shaw, 15, 27, 500, 25, 30
Samuel Stevens, 12, 4, 160, 2, 30
Henry Showalter, 50, 55, 1000, 10, 165
Alfred Efaw, 12, -, 120, 5, 160
John Cross, 6, 9, 190, 5, 40
James Park, 105, 119, 2500, 93, 208
John Paine, 80, 156, 2000, 100, 235

John Lemley, 10, -, 100, 2, 80
Charles Anderson, 50, 150, 2000, 75, 340
Harrison Anderson, 12, 48, 400, -, 125
B___ Postlewate, 16, 71, 400, 10, 120
___ham Morris, 40, 60, 1000, -, 100
_____ Pettigrew, 56, 354, 3400, 80, 345
John Hanes, 95, 84, 2000, 18, 256
_____ Hanes, 80, 45, 1500, 40, 610
Caleb Jackson, 80, 60, 1500, 15, 200
William Rush, 180, 200, 2500, 30, 378
_____ Delaney, 24, 76, 650, 15, 188
David Hendershot, 50, 183, 1400, 60, 232
Edmond Moore, 30, 100, 800, 10, 100
Anthony Morris, 75, 75, 1200, 50, 250
Isaac Lemasters, 40, 137, 1200, 10, 278
Gabriel Furbee, 65, 45, 1200, 40, 284
Wm. F. Clayton, 12, 6, 100, 10, 100
Robert Calvert, 40, 216, 1500, 10, 300
___ton Lemasters, 40, 113, 1200, 5, 250
Elizabeth Calvert, 35, 265, 1200, 10, 150
Asbury Marshall, 40, 260, 1200, 15, 150
Joseph Stewart, 14, 100, 1000, 10, 100
Jacob Sibol, 15, 65, 800, 10, 40
Charles Richmond, 10, 11, 150, 5, 25
Joseph E. Curier, 25, 95, 600, 10, 100
David S. Richmond, 45, 55, 800, 10, 124
John Simons, 45, 140, 1200, 75, 450
Joseph Rush, 40, 110, 1300, 10, 275
Sarah Rush, 40, 160, 2000, 5, 240

George C. Sirus, 50, 550, 3000, 5, 160

Absolom Postlewate, 100, 400, 3500, 60, 161

Joseph Postlewate, 11, -, 120, 10, 125

Nelson Postlewate, 30, 20, 500, 10, 294

Wm. Postlewate, 40, 80, 600, 15, 110

Barbara Higgins, 40, 60, 1000, 10, 314

Christopher Young, 80, 120, 1500, 12, 150

James Dunham, 30, 58, 1200, 60, 230

Joseph Sliper, 20, 72, 800, 10, 255

John Hurly, 20, 85, 800, 10, 110

George Folan, 50, 75, 1250, 60, 150

Nicholas Pratt, 15, 65, 700, 10, 105

Simon Mathews (Matheney), 20, 80, 800, 10, 125

John Chaves, 40, 80, 800, 15, 165

Samuel Postlewate, 20, 30, 200, 5, 30

Peter Hossick, 40, 35, 800, 15, 200

George Wickham, 11, 116, 600, 2, 125

Lucas Miller, 17, 33, 500, 10, 100

Lindsay Stuner (S. Turner), 40, 110, 1150, 40, 334

___al Hite, 30, 125, 1500, 5, 190

Samuel Furbey, 50, 50, 1100, 50, 234

John Furbey, 50, 50, 1000, 15, 150

Frederick Bookner, 20, 50, 700, 15, 125

Mary Bookner, 20, 51, 700, 5, 150

Jesse Goddard, 38, 48, 600, 12, 125

John Goddard, 35, 65, 1000, 45, 300

David Goddard, 14, 36, 300, 15, 100

George Homan, 15, 50, 400, 15, 150

Rezin Goddard, 40, 2, 400, 12, 100

Henry Hall, 9, -, 90, 5, 100

Harrison Yoho, 35, 61, 800, 40, 192

Wm. R. Goddard, 35, 5, 400, 5, 146

Samuel Arnold, 115, 459, 3953, 10, 253

Daniel Finch, 30, 46, 830, 3, 112

Danl. Finch, 30, 46, 820, 5, 100

Francis Bessard, 30, 30, 600, 10, 125

Sarah Loveage, 7, 78, 300, 5, 100

James Gainer, 30, 100, 1300, 5, 125

John Kirkpatrick, 35, -, 200, 5, 100

George Canfield, 80, 70, 2000, 40, 509

Madison Adis, 10, 20, 200, 40, 150

Nicholas Cooper, 25, 75, 100, 12, 200

Jenny Smith, 16, 109, 100, 60, 110

John Parsons, 200, 600, 800, 125, 1053

_. J. Metcalf, 30, 70, 900, 15, 120

John Q. Graham, 21, 96, 400, 6, 100

Robert Butler, 14, 36, 300, 5, 20

R. B. Butler, 20, 88, 700, 65, 150

Isaac Fulkerson, 7, 92, 600, 5, -

Jams Moser, 30, 66, 1000, 10, 120

Thomas Lowrey, 125, 200, 3000, 100, 567

Hezekiah Wayman, 40, 63, 1000, 80, 466

Samuel Earlywine, 25, 25, 600, 5, 125

Samuel Connor, 30, 45, 707, 10, 185

_____ Layhue, 50, 95, 1600, 24, 387

_____ C. Layhue, 5, -, 100, -, 40

Jacob Moore, 100, 100, 5000, 50, 258

Samuel Moore, 100, 100, 5000, 350, 834

Samuel Clark, 5, 6, 100, 5, 150

John Moore, 40, 120, 1500, 30, 1216

Isaac Smith, 50, 200, 2500, 100, 542

Nathan Allen, 70, 20, 2500, 100, 576

Gabriel Debolt, 20, 109, 750, 120

Josiah Hawkins, 20, 80, 1800, 10, 40

_____ McCloud, 8, 44, 520, 5, 437

Wm. H. Trader, 45, 105, 1800, 5, 113

Thos. Kirkpatrick, 20, 30, 300, 5, 25

Isaac Paugh, 45, 110, 1450, 15, 463

Elijah Hendershot, 7, 25, 250, 5, 112
George Yoho, 60, 60, 1080, 35, 377
_____ Hawkins, 7, 42, 200, 10, 100
George White, 30, 20, 500, 50, 100
J____ K. Kewins, 25, 25, 200, 10,
110
Benj. Johnson, 65, 26, 1200, 15, -
Levi Shuman, 80, 219, 2000, 40, 835
_____ Delaney, 80, 280, 2600, 40,
470
Wm. Delaney, 6, 12, 420, 5, 72
George Watson, 2, 100, 300, -, -
Oliver Brock, 16, 24, 150, 5, 40
Joseph Brock, 30, 30, 800, 5, 100
George Kiger, 40, 100, 840, 15, 196
James Guthrie, 75, -, 750, 12, -
Alex. Kirkland, 40, 48, 888, 15, 405
Joseph Taylor, 18, 50, 500, 40, 180
William Wade, 7, 93, 300, 5, -
John J. McGinnis, 10, 40, 300, 5,
140
James Wade, 15, 35, 250, 5, 160
John Jolliff, 20, 45, 600, 10, 25
Thomas Wade, 10, 100, 600, 10, 110
Henry Smith, 20, 45, 600, 5, 105
John T. Smith, 40, 60, 800, 20, 125
Thomas Vandine, 10, 5, 150, 5, 30
Hezekiah Owens, 35, 65, 800, 5, 120
Emanuel Vandine, 20, 25, 400, 10,
100
_____ Vandine, 15, 35, 300, -, 30
Frederick Adams, 60, 40, 1500, 20,
225
_____ Allen, 50, 70, 1200, 10, 123
_____ J. Morgan, 235, 40, 1500, 10,
100
_____ K. Niceley, 35, 61, 800, 16,
316
David Garlo, 45, 45, 880, 18, 214
_____ Rodgers, 100, 80, 2000, 60,
350
Allen Hill, 11, 15, 250, 5, 45
Elias Workman, 15, 35, 400, 3, 25
_____ W. Garner, 200, 700, 1500, 6,
908
John Yoho, 25, 65, 1080, 24, 210

John Higgins, 40, 132, 1700, 100,
400
John Hafer, 25, 53, 800, 24, 170
Elizabeth Dickey, 40, 30, 700, 7, 129
Edward Pharis, 150, 65, 3000, 100,
844
_____ Allen, 35, 89, 1200, 12, 250
Eli Anderson, 26, 100, 1000, 4, 110
J. W. Walker, 15, 50, 660, 9, 150
Franklin Hall, 75, 35, 1250, 75, 580
Solomon Blake, 50, 50, 1000, 20,
200
William Peg, 40, 44, 900, 50, 215
Samuel O'Neal, 30, 37, 174, 10, 215
William Hartley, 40, 60, 1200, 15,
210
John Burton, 25, 97, 800, 25, 343
James Stewart, 80, 82, 1622, 50, 430
James Baxter, 50, 105, 1920, 12, 244
Sarah Briggs, 35, 30, 660, 20, 175
Stephen Horner, 10, 20, 200, -, -
_____ Dunlap, 70, 60, 1260, 100,
417
Michael Huff, 35, 65, 800, 12, 254
_. V. Hider, 40, 36, 800, 12, 200
John Horner, 10, 15, 250, 10, 40
Aaron Haney, 50, 100, 2000, 15, 125
John Huff, 40, 80, 2000, 70, 600
Lazarus Trader, 15, 10, 250, 12, 208
Thomas Huff, 90, 145, 4000, 50, 327
Abram Simonson, 14, 86, 600, 10,
100
David Briggs, 80, 190, 3600, 150,
537
Robert Lemons, 27, 80, 1000, 40,
300
Joseph Powell, 50,-, 500, 15, 200
_____ Wright, 204, 80, 800, 15, 150
Johnathan Cox, 35, 200, 1600, 10,
285
_____ Allento, 60, 120, 2500, 15,
258
Wesley Leap, 100, 56, 2000, 15, 200
Winfield Robinson, 23, 34, 300, 10,
200

Eugenius Walker, 30, 10, 600, 40, 190

William Moore, 45, 80, 1000, 20, 129

John McCloud, 70, 47, 2166, 60, 890

Gabriel Jackson, 25, 50, 400, 10, 50

George Lockhart, 16, 36, 375, 16, 127

Gabriel Leap, 45, 55, 1600, 15, 250

Nelson Garner, 100, 100, 3000, 15, 262

John Bardine, 60, 140, 2000, 5, 20

William Haney, 50, -, 500, 12, 100

Joseph Cox, 25, 150, 1360, 30, 296

Nacer Clark, 275, 700, 800, 150, 655

Josephus Clark, 40, 70, 1000, 15, 200

Daniel Handlen, 20, 30, 400, 12, 185

Wesley Leap, 25, -, 250, 70, 112

Robert Leap, 100, 160, 3120, 200, 708

Francis Doran, 25, 88, 1200, 100, 316

George Franks, 80, 164, 1600, 75, 722

George Spiser, 30, -, 200, 5, 100

Thomas Burges, 65, 90, 1500, 75, 308

Edward Kerwen, 4, 16, 160, 10, -

Samuel Coats, 60, 64, 1000, 7, 176

F. M. Taylor, 30, 700, 1000, 8, 150

James Williams, 30, 33, 504, 15, 322

Isaac Coen (Corn), 60, 40, 1000, 15, -

Oliver Sloan, 24, 143, 1400, 5, 75

David Vandine, 10, 15, 250, 5, 110

Hugh Kirkland, 50, 50, 800, 50, 299

Daniel Montgomery, 18,-, 180, 10, 112

David Montgomery, 13, 32, 500, 10, 100

George Palmer, 60, 29, 623, 25, 240

Alfred Wilkinson, 20, 30, 500, 15, 155

William Allen, 40, 80, 800, 15, 150

John Cain, 10, 40, 400, 10, 100

Emanuel Amos, 46, 90, 2000, 25, 225

John Sant__, 35, 49, 500, 5, 145

James Coen, 30, 100, 480, 30, 125

George Amos, 22, -, 220, 10, 120

William Laflin, 100, 150, 2500, 75, 340

Joseph Bland, 25, 50, 500, 15, 254

Marcus Lemly, 25, 55, 400, 15, 105

Robert McAllister, 14, 36, 300, 15, 50

Isaac Cross, 8, 2, 80, 10, 100

John Owens, 25, 60, 500, 15, 283

Johnathan Nott, 15, 22, 75, 5, 40

George J. Steel, 75, 145, 2000, 125, 412

Frederick Steel, 80, 170, 2500, 30, 380

William Watson, 80, 52, 400, 10, 100

John Furbey, 60, 167, 2500, 100, 560

John Rice, 40, 400, 3000, 50, 408

Samuel Bland, 35, 63, 2500, 25, 607

David Bland, 40, 180, 1500, 12, 275

Allen Street, 24, -, 140, 10, 100

Richard Cook, 100, 312, 8240, 126, 960

Ira Cook, 26, 94, 1500, 12, 245

Thomas Steel, 150, 250, 6000, 100, 1076

Alfred B. Steel, 150, 250, 6000, 10, 205

Wm. Snodgrass, 24, 76, 400, 2, 75

Robert Guthrie, 30, 73, 800, 10, 116

John Guthrie, 6, 100, 800, -, 106

John C. Hart, 60, 45, 1000, 100, 300

Commodore P. Guthrie, 15, 117, 1100, 10, 25

Enoch Hawkins, 16, 43, 300, 15, 200

Albert Haycock, 20, 40, 400, 12, 700

F__ Pairo, 70, 175, 2000, 10, 333

George Snodgrass, 8, 69, 600, 15, 140

Wm. Guthrie, 35, 165, 2000, 30, 245

James Bickerton, 50, 70, 900, 35, 330

Elizabeth Buckhannon, 50, 50, 700, 12, 393

Indiana Tomlinson, 50, 50, 1000, 14, 106

Louis Lantz, 35, 65, 1200, 15, 258

Margaret Bar, 14, 12, 390, 10, 200

Van Buckhannon, 15, 85, 2000, 5, 112

Elijah Morgan, 16, 628, 6000, 15, 418

Morris Cussick, 20, 34, 500, 6, 172

Charles Swisher, 16, 32, 400, 10, 23

William Huff, 100, -, 6000, 50, 180

Aaron Morgan, 30, 40, 1200, 10, 250

Wm. A. Hitchcock, 12, 58, 200, 10, 112

Jakelous Boyer, 40, 141, 2000, 20, 338

F. W. Hitchcock, 15, 52, 500, -, 40

E. W. Hitchcock, 40, 53, 800, 10, 295

Mathias Whiteman, 40, 160, 1500, 25, 325

Marshall Whiteman, 25, 15, 800, 20, 84

Patric Slobul, 50, 25, 425, 10, 154

John Jennings, 50, 125, 1200, 10, 387

Thomas Noble, 15, 35, 400, 10, 112

Levi Cox, 10, 62, 400, 10, 110

John Snodgrass, 50, 250, 2000, 50, 402

Thomas H. Snodgrass, 20, 80, 400, 10, 103

Washington Snodgrass, 40, 78, 1000, 10, 420

Henry Fleharty, 7, 68, 400, 7, 117

Michael Eddy, 12, 63, 400, 2, 148

Elias Jones, 30, 14, 400, 25, 200

Jams Buckhannon, 35, 125, 2300, 25, 690

David Brooks, 20, 120, 2000, 15, 200

Henry Hossack, 50, 50, 2000, 10, 140

William Fogg, 15, 470, 1000, 10, 350

James Garnett, 4, 20, 180, 5, 125

Elizabeth Moore, 50, 110, 4400, 30, 250

Thomas Snodgrass, 19, 125, 1000, 20, 275

Colamore Grose, 40, 60, 500, 15, 222

Adam Stone, 25, 100, 901, 15, 125

Nelson Miner, 22, 100, 900, 15, 125

John Shuman, 40, 65, 1050, 25, 200

Hamilton Inman, 35, 140, 1500, 5, 100

Thomas Maple, 40, 164, 1000, 25, 160

Samuel Goddard, 95, 100, 2500, 150, 470

Samuel Stooley, 18, 35, 300, 3, 67

John Lap, 45, 66, 1500, 20, 407

Mary Cole, 20, 113, 1300, 5, 100

Jacob Hindman, 20, 34, 500, 40, 160

Jacob Hareflinger, 12, 102, 600, 13, 173

Wm. Jennings, 15, 60, 750, 10, 100

Johnathan Potts, 18, 100, 1000, 5, 150

Samuel Hawkins, 15, 15, 150, 5, 40

Victor Borrer, 18, 330, 2000, 5, 105

Hiram Hanes, 60, 440, 10000, 100, 380

Daniel Huff, 30, -, 1200, 15, 222

Joseph Hall, 160, 314, 18800, 125, 800

_____ Powell, 86, -, 320, 10, 170

Robert Ligget, 120, 47, 6000, 10, 585

Josiah Cunningham, 115, 53, 6000, 125, 440

_____ Thistle, 200, 1100, 30000, 125, 1020

John B. Brown, 100, 300, 8800, 100, 250

Elijah Potts, 10, -, 400, 10, 25

Thomas Williams, 100, 257, 8000, 100, 888

Andrew McEldowney, 150, 70, 10200, 60, 594

Nicholas Bandy, 55, -, 2000, 100, 200

Samuel McEldowney, 84, 86, 10000, 100, 860

Anthony Loveall, 100, -, 4000, 10, 250

Jeremiah Loveall, 43, -, 1720, 25, 456

Stephen Loveall, 15, -, 600, 45, 114

James Shuman, 40, 53, 1200, 15, 246

John Travis, 40, 146, 4000, 50, 875

Preston Crump, 36, 30, 1000, 10, 125

Benedict Enoch, 23, 27, 800, 40, 140

John Smith, 35, 55, 1000, 10, 100

John Probest, 18, 145, 1500, 15, 120

Peter Travis, 6, 100, 600, 15, 160

_. W. Corothers, 35, 45, 1200, 20, 250

Alex. Van Camp, 40, 60, 1500, 25, 203

John Christian, 35, 65, 1200, 30, 150

James Henry, 15, 25, 400, 10, 115

Rotla Van Camp, 25, 29, 1000, 10, 124

Solomore Grim 14, 62, 500, 10, 135

James Glenn, 30, 70, 1500, 40, 340

James Ellet, 40, 90, 1600, 5, 100

John Bruner, 9, 81, 800, 5, -

Butler Van Camp, 45, 60, 2000, 90, 219

John Van Camp, 50, 90, 4000, 152, 457

Mavis Meredith, 50, 45, 2000, 30, 314

David Jacobs, 20, 30, 400, 10, 200

James Buckhannon, 40, 11, 1000, 15, 175

Reuben Martin, 30, 35, 1200, 5, 170

_____ Van Camp, 30, 165, 2340, 100, 493

John Hickman, 20, 83, 1200, 5, 40

_. B. Jenkins, 40, 105, 1740, 60, 310

_____ Venoni, 15, 38, 795, 10, 205

_. H. Evans, 30, 43, 800, 20, 258

_. F. Van Camp, 70, 89, 1800, 60, 272

Joseph Margus, 38, -, 380, 10, 110

Bobbit Fulks, 14, 86, 1000, 40, 125

Charles Fulks, 5, -, 300, 10, 30

Jacob Boner (Borrer), 16, 84, 700, 60, 240

Alfred Bolin, 7, 3, 100, 5, 100

George DeBrick (Detrick), 20, 106, 1100, 10, 150

Henry Minemiar, 12, 37, 500, 10, 150

William Price, 50, -, 600, 75, 400

Samuel Dennis, 50, 50, 2000, 75, 350

Jesse Patton, 100, 250, 10000, 100, 575

Joseph Patton, 170, 170, 10000, 150, 900

David Skinner, 25, 15, 800, 10, 230

R. W. Lock, 30, 25, 1600, 3, 30

John McCoskey, 50, -, 2500, -, 187

William Gorby, 54, 1, 2000, 60, 150

William Palmer, 28, 25, 700, 100, 400

Philip Whitten, 70, 75, 6000, 125, 625

R. W. Cox, 220, 580, 15000, 100, 575

John J. Yarnall, 140, 150, 10000, -, 110

Leroy R. Wise, 10, 990, 5000, 20, 186

Friend Cox, 42, -, 3500, 2, 10

B___ F. Martin, 45, 55, 2000, 120, 535

Joseph Boyers, 100, 260, 10000, 150, 1000

Joseph Whiteman, 25, 125, 1600, 25, 346

Wm. A. Newman, 40, 100, 1600, 25, 300

George Bland, 60, 223, 2600, 180, 503

Wirt County, West Virginia
1860 Agricultural Census

The University of North Carolina at Chapel Hill filmed the 1860 agricultural census for Wirt County from originals at the West Virginia State Archives under a grant from the National Science Foundation in 1963.

Columns 1, 2, 3, 4, 5, and 13 represent the following information on the census:
1. Name of Owner, Agent or Manager of Farm
2. Acres of Improved Land
3. Acres of Unimproved Land
4. Cash Value of the Farm
5. Value of Farming Implements and Machinery
13. Value of Livestock

This county had a large number of tenants.

Samuel C. Morehead, 60, 20, 1200, 10, 305
G. W. Lockhart, tenant, -, -, -, 111
John M. Lockhart, 70, 712, 3208, 202, 464
Albert Lockhart, 8, 79, 300, -, 86
John Ott, 80, 128, 1500, 100, 386
Alfred Ott, 60, 103, 1200, 5, 412
Tidillis Ott, 60, 165, 1800, 45, 423
Tidillis Ott Jr., -, 150, 600, -, 100
Nimrod Wiseman, 150, 700, 4198, 75, 803
Wilson McClung, 3, 110, 339, 3, 106
D. & A. Thomas, 10, 15, 400, 3, 129
John Dobson, tenant, -, -, -, 75
Lewis Sheppard, 60, 106, 1666, 40, 535
David Enoch, 16, 186, 500, 6, 40
Samuel Sheppard, 120, 130, 3000, 100, 657
Jothathan Sheppard, 100, 1600, 3000, 50, 892
John Thorn, 25, 105, 500, 5, 215
William O. B. Sheppard, 35, 80, 600, 5, 96
Thomas Thorn Jr., 35, 90, 1000, 60, 147
Michael Thorn, 70, 130, 1000, 50, 446

Thomas Thorn Sr., 25, 100, 800, 5, 164
William H. Thorn, 80, 150, 1500, 125, 507
Eugene Thorn, 40, 278, 636, 10, 286
Joseph Pickergill, 40, 164, 500, -, 16
Michael Herfman, 60, 140, 800, 75, 188
Nathaniel Herfman, 8, 142, 500, 2, 75
John Elliott, 15, 60, 300, 3, 90
W. H. Rexrode, tenant, -, -, 5, 68
Jefferson Ayres, 20, 65, 400, 5, 125
Johnathan Sheppard, 40, 100, 500, 40, 328
Wilson Sheppard, 50, 193, 1000, 40, 241
Jacob Bumgarner, 100, 300, 4000, 150, 1334
David Summervill, 75, 195, 1500, 75, 237
Alfred Sumervill, 75, 475, 2500, 100, 590
S. H. Morehead, 50, 250, 1150, 100, 528
Ester McClung, 6, 194, 450, 1, 63
William _. Mcfee, 25, 225, 500, 10, 279
R. G. Sheppard, tenant, -, -, 40, 298

Alexander Hestler, 30, 425, 1800, 10, 225

Sherley Enoch, tenant, -, -, -, 130

William Sheppard, 400, 3000, 15000, 250, 1720

Martin Sheppard, 25, 75, 500, 8, 142

John Anderson, 14, 137, 600, 5, 96

Joseph Grim, 30, 75, 530, -, -

Daniel Caine, 30, 70, 2000, 10, 128

Henry Enoch, 60, 400, 2000, 10, 227

Leonard McTran, tenant, -, -, -, 73

James Clarke, -, 650, 1000, 10, 80

Minyard Harris, 60, 142, 1800, 8, 124

John Harris, 30, 370, 1200, 10, 76

Reuben Full, 100, 290, 1000, 15, 389

Samuel Sheppard Jr., 34, 130, 1000, 5, 136

Joseph Lawson, 10, 40, 200, 2, 53

Catharine Dobson, 25, 125, 750, -, 31

Abraham Williams, 12, 63, 200, 3, 20

J. W. Shaver, 20, 55, 300, 10, 86

Wilson Riley, tenant, -, -, 5, 121

William F. Candle, 20, 100, 300, 2, 120

William T. Bates, 20, 86, 600, 20, 127

William George, 35, 133, 147, 5, 100

Samuel Wagner, 15, 35, 225, -, 13

William Collison, 75, 68, 1716, 170, 692

John Smith, 130, 240, 4000, 35, 204

Ephraim T. Doolitel, 12, 117, 300, 5, 26

William Faylor(Taylor), tenant, -, -, -, 38

William N. Smith, tenant, -, -, -, 15

John L. Boggs, 125, 375, 5000, 75, 553

Thomas Lev Jr., 60, 50, 1000, 50, 233

Peter Conrad, 100, 50, 3000, 75, 800

Jacob Conrad, 25, 25, 300, 10, 114

John Conrad, 25, 50, 200, 3, 120

Lewis Eye, tenant, -, -, -, 50

George C. Seman, 40, 476, 648, 5, 96

Charles Boggs, tenant, -, -, 7, 181

William McCutchen, 35, 150, 1000, 80, 300

Andrew Colison, 25, 178, 800, 12, 133

Harrison McCutchen, 6, 70, 400, 5, 170

J. B. Hilbert, 30, 120, 500, 3, 114

Elisha McCutchen, 5, 145, 400, 5, 244

James McCutchen, 12, 100, 400, 5, 44

John B. Lee, tenant, -, -, 5, 38

Lanias Lee, 100, 90, 2500, 10, 154

William Lee Jr., tenant, -, -, -, 243

Elisha Baker, 6, 700, 1300, 8, 253

Joseph Malcom, 80, 30, 1400, 72, 342

Samuel Malcom, tenant, -, -, 5, 24

Zedock Thorn, 80, 73, 3000, 100, 500

Henry Amick, 120, 130, 3700, 10, 480

John _. Burchett (Burdett), tenant, -, -, -, 105

Jesse Thornton, tenant, -, -, -, 208

John Howard, 15, 35, 300, -, 57

William Cain, tenant, -, -, -, 91

Thornton Baker, 50, 275, 2000, 10, 172

Samuel Thornton, 70, 144, 1800, 100, 196

John Thornton, tenant, -, -, -, 22

Nathaniel Morehead, 3, 90, 300, 5, 120

John Pitsenbarger, tenant, -, -, -, 130

(Thomas) A. Paxton, 75, 104, 2000, 10, 334

A. P. Fought, 100, 243, 3000, 80, 944

John Norris, 18, 95, 200, 5, 31

Alfred Lott, 3, 17, 100, 5, 174

Presley Morehead, tenant, -, -, -, 64

Thomas J. Morehead, 50, 185, 1500, 10, 64

Thomas B. McFarland, 50, 176, 1356, 15, 162

David Hopkins, 100, 25, 3000, 100, 270

Edward Knotts, 100, 225, 2000, 10, 400

Henry Lockhart, 10, 105, 450, 2, 125

Levi Hopkins, tenant, -, -, -, 164

George Dent, 125, 250, 5300, 100, 695

William Vansickles, 190, 30, 4000, 100, 476

Alen Fricket, 75, 25, 1500, 100, 491

A. J. Hickman, -, -, -, 10, 130

C. W. Rogers, 85, 4915, 9000, 75, 492

John Rogers, -, -, -, -, -

Ephraim Sayres, 60, 44, 1500, 5, 42

Ely Judson, 20, 27, 150, -, 58

Dissnay Dye, 30, 70, 300, 12, 218

Samuel Bonnett, 10, 25, 250, 10, 43

Lewis C. Coe, 100, 175, 2500, 250, 625

Levi Coe, 20, 147, 800, 10, 126

Asa Wilson, tenant, -, -, -, 30

Robert Monroe, 60, 60, 800, 5, 152

John Hall, 3, -, 800, -, 16

James Monroe, 30, 30, 1500, 15, 317

Elisha Hickman, 30, 60, 1500, 5, 123

William Hickman, tenant, -, -, 5, 163

George Cline, 30, 1600, 1600, 5, 111

Samuel Lee, 7, 57, 500, 2, 69

Joshua Lee, 50, 500, 1000, 80, 245

King Berry Dulin, 20, 80, 700, 5, 93

Jams White, 9, 5, 600, -, 51

Sidney Enoch, 6, 34, 500, 5, 40

Samuel Davis, 35, 89, 500, 5, 146

Rebecca Corbett, 25, 275, 500, 5, 57

George Peck, 12, 38, 250, 5, 74

Nathaniel Boggs, 40, 160, 400, 4, 154

Harriet Boggs, 5, 45, 100, 3, 117

A. Perrill, 2, 100, 200, -, -

Henderson Petty, 12, 163, 500, 2, 65

Thomas Boggs, 6, 75, 250, 60, 145

Susan McGlothin, 25, 75, 300, 5, 88

John F. Petty, 60, 60, 1500, 50, 232

Richard Petty, 20, 55, 800, -, 203

Elias Bennett, tenant, -, -, -, 23

Thomas R. Parker, tenant, -, -, 5, 98

John A. Farfax, tenant, -, -, 5, 96

Enoch Deamer, tenant, -, -, 15, 235

D. A. Sayers, 18, 32, 600, -, -

William Dulin, 30, 110, 800, 5, 98

James Petty, 30, 20, 1000, 10, 167

A. Beauty, 200, 300, 5000, 125, 746

Hanibal McClaine, 100, 174, 2000, -, -

Cornelius Wayne, 10, 172, 50, 5, 40

John C. Rathbone, 150, 800, 10000, 67, 595

Alfred Scott, 25, 75, 700, 20, 154

John Cehorn, -, -, -, -, 100

Thomas Henderson, 200, 300, 4000, 30, 17

Nathaniel McDonnold, -, -, -, 18, 165

Jacob McDonnold, -, -, -, -, 144

John Collins, -, -, -, -, 16

Rowland Petty, 100, 150, 1200, 10, 449

Luther Owens, 7, 56, 300, 20, 168

Willis Owens, 150, 75, 1500, 8, 90

Sandy Owens, tenant, -, -, 5, 185

B. F. Renolds, 40, 79, 600, 30, 256

Parish Bailey, -, -, -, -, 37

James Deaver, 40, 70, 800, 20, 195

Samuel Mcfee, 50, 75, 1000, 25, 200

John Mcfee, 15, 99, 400, 5, 121

Samuel Edwards, 20, 408, 700, 30, 367

John Weaver, 6, 108, 114, 3, 95

Mary Edwards, 25, 30, 350, 5, 63

Richard Grim, tenant, -, -, 10, 136

R. P. Morgan, 100, 80, 1600, 65, 197

Colintine Cain, 40, 41, 500, 10, 234

James Cox, tenant, -, -, 5, 93

Alfred Cockhern (Cockbern), 12, 128, 450, 5, 22

A. House, 30, 184, 700, 8, 148

Henry Hudson, 4, 48, 150, -, 48

Henry Buckley, 20, 180, 800, 4, 100
Henry Lowther, 80, 260, 1200, 15, 136
William Bailey, 60, 40, 1000, 144, 265
John A. Lowther, 5, 60, 275, 2, 25
Alen Hendershot, tenant, -, -, -, 40
William Law, 12, 55, 300, -, 50
Jacob Bailey, tenant, -, -, 40, 215
Obedial Hendershot, 30, 42, 600, 30, 58
William Bailey Sr., 100, 165, 2000, -, 85
Samuel Bailey, 17, 43, 300, 3, 87
R. W. Hammond, 80, 150, 4000, 50, 414
C. W. Rockhold, 15, 5, 250, 7, 40
Minyard Rockhold, 20, 3, 1200, 30, 158
Isaih Favenner, 250, 350, 8100, 75, 675
Jeffery Wiseman, 80, 100, 2000, 40, 522
Joseph Collins, 20, 80, 200, 4, 50
L. C. Rogers, tenant, -, -, 5, 150
Isaih Ruble, 50, 150, 2000, 25, 70
Joseph Ott, 50, 150, 1200, 50, 181
Lewis Ott, -, -, -, 3, 58
Michael Thorn Sr., 50, 130, 1000, 20, 192
Caleb Wiseman, 50, 250, 1300, 40, 360
John Lockhart, tenant, -, -, 3, 74
Robert Wallis, 70, 180, 1200, 15, 341
Gilead Lockhart, 10, 130, 200, 5, 62
Elizabeth Woodyard, 40, 75, 800, 8, 235
William Lockhart, 35, 260, 1200, 10, 111
Dillis Woodyard, 15, 145, 480, 8, 102
Jaramah Woodyard, 75, 145, 500, 5, 217
David Horner, -, 50, 50, 3, 231
Mathew Edminson, 1, 110, 600, 4, -

E. T. Vaught, 15, 60, 150, 1, 81
John Walters, 4, 71, 130, 15, 125
Morris Sprouse, 30, 270, 800, 5, 56
John D. Beleard, -, 50, 250, 25, 124
Samuel Bowie, 40, 460, 1200, 30, 88
Samuel Martin, 20, 130, 800, 4, 111
Michael Dolin, 100, 250, 1000, 15, 245
Israel Banner, 45, 255, 1500, 60, 246
John Winer, 40, 100, 1000, 10, 227
John Jacobs, 30, 50, 600, 5, 94
Jacob Brookever, tenant, -, -, -, 12
George L. Daggoth, 25, 55, 400, 4, 74
Elkana Lions, 20, 45, 400, 5, 70
E. G. Coe, 20, 82, 1000, 10, 142
James P. Ball, 50, 95, 1500, 15, 200
George Stephens, 50, 120, 3000, 75, 365
H. D. Richards, 60, 420, 2000, 75, 404
Thomas Ruble, tenant, -, -, -, 71
George C. James, 20, 360, 900, 10, 127
G. W. Davis, 15, 185, 00, 3, 94
Johnathan Bennett, 5, 45, 150, -, -
James D. Gates, 25, 302, 1000, 2, 256
Jams W. Morehead, 30, 325, 1000, 5, 140
Abraham Plum, 9, 241, 600, 3, 14
William W. Lions, 10, 68, 300, 5, 141
Jarrett B. Ruble, 3, 95, 300, -, -
Abraham Vaught (Haught), 25, 125, 800, 5, 61
Elkana Bennett, 25, 100, 500, 40, 123
JamesWilson, 20, 30, 300, 5, 137
John Masters, 35, 115, 800, 20, 86
Peter Martin, 30, 300, 400, -, 47
Arnold Bennett, 70, 200, 1500, 75, 316
Thompson Gates, 40, 500, 1500, -, 214

Joseph Bouise, 50, 150, 1500, 40, 294

Michael Halbert, 2, 68, 200, -, 24

James M. Lions, tenant, -, -, 2, 185

Askuck Roberts, 20, 100, 400, 30, 175

Marthy M. Ball, 20, 100, 1000, -, 38

William Starke, 18, 106, 500, 3, 108

James Lumness, tenant, -, -, 5, 37

Mary Ball, 40, 260, 1500, -, 55

B. F. Ball, 30, -, 1200, 60, 326

Louisa F. Ball, 70, -, 600, -, -

Layfitt Ball, 25, -, 600, -, 100

William L. Ball, 25, 81, 1500, 20, 173

James Wilson, tenant, -, -, -, 25

Lewis Woodyard, 130, -, 5000, 100, 616

Peter Spoon, 3, 97, 100, -, -

Michael Hanaman, 50, 265, 1200, 50, 382

Henry Kernes, 10, 100, 400, -, 45

Sandy Sprouse, tenant, -, -, 15, 110

Solomon Mulinaxe, 7, 60, 75, 3, 123

James McDonnol, tenant, -, -, 5, 85

Richard Timms, 140, 140, 2000, 70, 271

William F. Petty, 80, 70, 600, 50, 184

Morton Baker, 40, 240, 1200, 5, 179

H. T. Wagel, 25, 87, 700, 5, 158

George Snyder, 70, 60, 1600, 30, 150

Jane Steel, 65, 35, 3000, 75, 733

R. P. Steel (Steed), 107, 1335, 5000, -, -

Frederick Bailey, 75, 50, 1000, 10, 200

James Baker, 24, 70, 1200, 20, 51

Charles Bailey, tenant, -, -, 5, 100

Franklin Hannon, 60, 20, 2000, 75, 324

Stanly Robinson, 30, 90, 1600, 10, 172

William Timms, 110, 60, 2000, 60, 520

Francis Barnes, 40, 80, 1000, 5, 141

James Nash, 30, 20, 600, 8, 103

George Jaco, 20, -, 250, 40, 295

C. W. West, 50, 70, 1400, 60, 195

Haille M. Creel, 35, 104, 1400, 5, 88

Sylvanius Barnes, 50, 70, 1000, 10, 63

J. W. Crawford, -, 400, 400, 5, 56

Harvey Bailey, tenant, -, -, -, 79

C. W. Braiden, 60, 50, 1000, 5, 186

Hugh P. Barnes, 70, 23, 900, 10, 184

Benjamin Roberts, 150, 50, 2500, 40, 718

Stephen Merrill, 120, 90, 300, 75, 523

William C. Wells, 80, 160, 2000, 75, 540

R. S. Brown, 40, 160, 2725, -, -

A. Beauchamp, 85, 15, -, -, -

James Fisher, tenant, -, -, 50, 416

Levi Wells, -, 30, 100, -, -

C. Rathbone, 75, 275, 1000, 10, 125

E. C. Hopkins, 30, 500, 2400, 60, 224

Jesse Lee, 11, 25, 600, -, 30

James Robinson, 100, 138, 4000, -, -

William Wells, 30, 30, 900, -, -

L. D. Woodyard, 20, 2, 900, 60, 258

Samuel Rathbone, 60, 430, 2500, 60, 191

Harrison Owens, 40, 160, 1200, 20, 223

E. S. Lockhart, 25, 175, 800, 5, 91

Caleb Lockhart, 24, 260, 800, 5, 122

Ely Robinson, 15, 70, 250, 5, 150

John Robinson, 40, 100, 800, 10, 137

Owen J. Morehead, 75, 8, 1200, 20, 176

Alfred Owens, -, 480, 480, -, 180

David Deavers, 100, 200, 2000, 100, 800

Jackson Deavers, tenant, -, -, 5, 350

Andrew Jones, 5, 45, 200, 1, 26

Cornelius Dent, 8, 62, 200, 3, 125

George Gray, 8, 62, 200, 2, 43

C. W. Fisher, 50, 90, 1200, 50, 415

C. R. Morgan, 15, 135, 450, 5, 104

Jessee Morgan, 40, 10, 400, 20, 194

Franklin Morgan, 20, 640, 1400, 2, 45

Stephen Ivins, 8, 42, 300, 4, 61

Clem Morgan, 20, 0, 200, 1, 122

Edmond Gatheny (Gothing), 15, 35, 350, 6, 17

Ivy Haught, 15, 185, 600, 5, 111

Josiah Ferrer, 40, 105, 800, 15, 153

Alpheus Dent, 60, 540, 1200, 14, 214

A. J. Roberts, 15, 187, 400, 5, 112

William Brummage, 18, 82, 500, 6, 123

Abner Pepper, 18, 132, 1500, 3, 27

Levi Brummage, 30, 600, 630, 60, 214

Isaiah Brummage, 10, 90, 200, 5, 50

Rowland P. Brummage, 4, 96, 200, 1, 69

Absolom R. Davis, 13, 87, 200, 3, 26

Jacob Wirt, 20, 310, 800, 5, 146

Peter Haught, 70, 250, 1200, 70, 268

Ely Wilson, 75, 325, 1500, 75, 324

Walter Bell, 15, 150, 500, 5, 83

Joshua Evins, 15, 285, 1000, 7, 53

George Monroe, 10, 90, 400, 4, 100

Sylus Bett, 75, 100, 1500, 35, 465

Marge Roberts, 75, 125, 1500, 100, 147

George W. Booker, tenant, -, -, 5, 133

Johnathan P. Booker, tenant, -, -, -, 86

Alandria Monroe, 165, 60, 150, 5, 112

William Fairfax, 90, 500, 3000, 20, 182

Bailey Knight, 30, 220, 1000, 5, 120

James J. West, 35, 365, 1200, 40, 15

Thomas Haught, 4, 96, 350, 2, 48

Amos Roberts, 50, 50, 500, 5, 168

George Mays, 3, 277, 75, 1, 68

Henry Dulin, 25, 115, 300, 4, 140

William F. Martin, 30, 270, 700, 5, 144

Seth Knight, 30, 27, 500, 5, 121

James H. Robinson, 20, 70, 200, 3, 65

Danuel Ledson, tenant, -, -, 1, 18

Gilmore Dye, 5, 62, 250, -, 38

Sanly Petty, 40, 100, 600, 5, 200

Samuel A. Petty, -, -, -, -, 25

John Badgely, 30, 120, 800, -, 25

Archibald Depugh, 40, 60, 1200, 10, 246

Hiram Depugh, 35, 65, 1200, 10, 246

Thomas Merrill, 15, 135, 600, 50, 199

William D. Beverlin, tenant, -, -, 5, 15

William H. Conrad, tenant, -, -, -, 156

Sylus Davis, 10, 90, 800, 3, 168

Thomas Price, 10, 130, 800, -, 24

Amos Price, 2, 8, 75, -, 20

George Miller, tenant, -, -, -, 287

Richard Bridges, tenant, -, -, -, 247

David Depugh, 100, 100, 2000, 5, 87

F. M. Hopkins, 35, 44, 1250, 1, 219

Bona Candle, 35, 44, 800, 35, 190

Hiram Pervel, 70, 30, 1500, 50, 287

Henry Miller, 25, 66, 1000, 6, 118

G. F. Petty, 15, 67, 500, 7, 79

William Merrill, tenant, -, -, 4, 74

Jackson Merrill, tenant, -, -, -, 37

John Hanner, 30, 130, 1200, -, 132

Joshua Benner, 25, 85, 600, -, 134

John Houchen, 40, 100, 2000, 50, 370

Marlin (Martin) Parker, 6, 94, 500, 10, 149

George W. McGinnis, 20, 87, 1000, 8, 165

Franklin Dulin, tenant, -, -, 5, 85

William Houchen, tenant, -, -, -, 75

Wesly Bufington, 30, 70, 1500, 10, 80

William Petty, 50, 250, 2000, 50, 352

C. R. Pickering, 8, 82, 360, 4, 70

Costiller Rathbone, 50, 2000, 4000, 60, 368

William Rathbone, -, 10000, 20000, -, -

Enoch Petty, 75, 132, 2200, 15, 415

Richard Petty, 24, 46, 1000, 10, 238

Z. W. Hickman, 41, 15, 800, 20, 83

William H. Foughty, 16, 99, 400, 5, 200

John T. Booker, tenant, -, -, -, 46

James Whitescotton, tenant, -, -, -, 66

Desey Dye, 40, 103, 1000, 40, 179

William Houchen, 25, 25, 600, 20, 233

William Houchen, 3, 45, 100, -, 24

Perry Houchen, 25, 25, 500, -, 35

H. H. Burner, 4, 46, 125, 10, 112

Aaron Ruble, 50, 170, 2000, 20, 368

William Cisson, 17, 33, 200, 4, 123

Boles Pickersman, 25, 275, 800, 5, 180

Charls Booher (Booker), tenant, -, -, -, 44

William Robinson, 30, 246, 1000, 8, 118

John Robinson, 45, 193, 2000, 10, 137

James Robinson, 25, 149, 700, 5, 183

Samuel Robinson, 10, 400, 800, -, -

Moses Marsh, 3, 750, 750, 4, 100

John Marsh, 3, 750, 750, 4, 10

David Watson, 12, 61, 500, 2, 131

Arish Garnson, 10, 98, 300, 2, 50

Anthony Harris, 25, 75, 800, 10, 183

Ruben Garnson, 5, 95, 150, 50, 47

Johnson Parrish, 20, 80, 600, 3, 46

William Parrish, 40, 60, 600, 5, 58

John Wilson, 12, 122, 400, 2, 94

R. M. Hall, 8, 32, 300, 3, 108

Thomas Baily, 7, 30, 300, 5, 157

James Robinson, 2, 98, 200, -, -

John Rockhold, 50, 300, 1000, 10, 175

John T. Robinson, tenant, -, -, -, 25

George Rockhold, 8, 92, 400, -, 85

L. R. Cox, 35, 76, 600, 60, 83

Wilson Bailis, 40, 10, 500, 4, 91

Ostin Bailis, tenant, -, -, -, 91

Peter Stutter, 16, 104, 500, 3, 76

William W. Foughty, 40, 110, 750, 10, 412

David Nott, tenant, -, -, -, 95

Simon Stutter, 5, 45, 100, 3, 12

David Cozall, 90, 267, 1550, 5, 149

Abraham Fought, 70, 102, 1500, 20, 376

Garey Hickman, tenant, -, -, 5, 16

Wilson Hickman, 30, 530, 1216, 5, 45

Solomon Foughty, 70, 40, 1500, 100, 413

N. A. Foughty, 40, 185, 800, 5, 123

Uriah Liason tenant, -, -, 5, 58

Mary Baily, 30, 170, 700, 5, 71

Daniel Fought, 41, 140, 1000, 100, 265

Jane Fought, 80, 6, 1500, 10, 230

Catharine Foughty, 50, 70, 1000, 10, 333

Thomas Fought, 130, 170, 2000, 105, 235

Adam Foughty, 75, 200, 2300, 25, 308

Thomas Peckison, 37, 87, 2200, 60, 333

J. N. Hail, 30, 75, 1200, 100, 307

Leonard Bidwell, 30, 75, 1200, -, -

John Parmentions, 60, 40, 800, 50, 72

A. Hally, 80, 160, 2200, -, 51

Amy Foster, 72, -, 2500, 50, 650

Fredrick Hunington, 3, -, 150, -, 53

Hedgeman Prible, 55, 30, 2000, 40, 453

Henry Prible, 55, 30, 2000, 40, 453

Benjamin Mound, 60, 50, 1500, 90, 460

A. Darnell, 63, 20, 1562, 125, 261

Wilson Bailey, tenant, -, -, -, 94

William Buffington, 80, 569, 2369, 40, 158

Jane H. Stud, tenant, -, -, -, 28

Martin Howly, 15, 100, 300, -, 21

Alfred Fought, 150, 450, 5000, 150, 566

Esn. Leap, 50, 79, 1000, 12, 81

Hiram Anderson, 21, 100, 500, 12, 81

Jefferson Founds, 3, 119, 350, -, -

John Leap, 60, 47, 949, 75, 269

Emry Leap, 15, 25, 300, 10, 202

John Leap Sr., 40, 27, 800, 5, 110

Samuel Leap, 60, 52, 1000, 20, 144

C. H. Cain, 40, 133, 1000, 16, 205

George Foughty, 100, 65, 2800, 20, 248

James Cain, 40, 25, 600, 75, 276

Elzy Leap, 20, 20, 400, 5, 109

Davis J. Deames, 15, 105, 350, 3, 35

William Wilson, 40, 352, 1375, -, 112

John Eppline, 14, 114, 600, 3, 98

Alfred Bell, 5, 117, 460, 10, 14

Juehat Higgins, 10, 90, 200, 10, 48

Henry Merrell, 1, 47, 100, 3, 32

Daniel Pickering, 100, 999, 3644, 38, 495

William Perrell, 25, 955, 1200, -, 13

Tompson Copen, 30, 170, 700, 10, 157

John Copen, 35, 135, 700, -, -

George Copen, 7, 493, 250, 5, 109

Layfitt Pickering, 2, 48, 200, 30, 400

Tompson Copen (Cosen), 15, 80, 500, 5, 130

Hiram Prible, 15, 6, 400, 20, 120

Martin Enoch, tenant, -, -, 20, 177

Isaac Enoch, 27, 25, 1000, 8, 121

Thomas Melvin, 50, 142, 1600, 20, 78

Daniel Woolf, 30, 1500, 2500, 15, 30

Layman Frazure, tenant, -, -, -, 87

___don Enoch, tenant, -, -, -, 183

Abraham Enoch, 250, 2350, 8000, 100, 1045

Andre Lyda, 25, -, 800, -, 100

Simeon Thornton, 150, 364, 4500, 60, 440

Lemuel Robinson, -, -, -, -, 86

Jessee Roach, 60, 140, 2000, 100, 271

Samuel Braidin, -, -, -, -, 26

A. H. Turley (Turbey), 35, 65, 1000, 15, 88

Abraham P. Enoch Jr., 65, 35, 1500, 10, 137

John Deames, tenant, -, -, 5, 85

William Lee, tenant, -, -, -, 121

Mathew Wilson, 12, 25, 600, 5, 100

John Merrill, 60, 150, 1200, 8, 195

Benjamin Robert, 50, 50, 1000, 40, 248

Isaac P. Wilson, tenant, -, -, 30, 185

John B. Ravenscroft, tenant, -, -, 30, 142

John W. Roberts, 40, 60, 700, 5, 127

Samuel Boothe, 35, 57, 800, 1, 117

Samuel Boothe, tenant, -, -, -, 173

Benjamin Rockhold, 10, 137, 250, 5, 81

John Cain, tenant, -, -, -, 24

James Harris, 30, -, 1000, -, 38

Butler Higgins, 50, 60, 1000, 25, 231

David Higgins, 25, 25, 1000, 75, 161

John Arust, 60, 17, 1000, 75, 204

Samuel Collums, 60, 90, 2000, 5, 161

Thomas Deames, 80, 170, 1125, 100, 400

Joshua Perrin, 25, 215, 1200, 25, 187

R. B. Perrin, 51, 195, 1564, 6, 187

John Thrash, 40, 160, 100, 50, 181

John Williams, 10, 90, 400, 2, 65

David Thornton, 80, 150, 2000, 8, 255

Nepth Deames, tenant, -, -, -, 20

George Creel, 25, 975, 20000, 150, 255

Isaac Nelson Jr., 20, 244, 400, 8, 176

Jesse Nelson, 70, 200, 2000, 40, 270

John Hall, 75, 255, 1500, 25, 236

James Nutter, tenant, -, -, -, 19
James H. Hall, 10, 108, 150, 4, 192
Umphry Nutter, 18, 22, 150, 8, 106
Jane Duff, tenant, -, -, -, 111
John Davis, 43, 454, 1197, 95, 311
G. H. Lemmons, 14, 356, 340, 5, 185
Benjamin Vernon, 120, 90, 4000, 85, 356
Morgan Pomroy, tenant, -, -, 10, 275
W. D. Wilkinson, 65, 75, 1500, 100, 900
Mason Certney, 50, 187, 800, 70, 54
Thomas King, 12, 81, 200, 3, 104
Nathaniel Morehead, 75, 55, 1500, 75, 419

James Stephens, tenant, -, -, 2, 39
Owen J. Coe, 50, 61, 600, 5, 100
Thomas Meridith, tenant, -, -, -, 15
Minie Newman, 50, 50, 3000, 172, 1780
Danuel Wilkinson, 100, 300, 4000, -, -
Heath Rockhold, 15, 485, 1000, 75, 287
Jane Nutter, tenant, -, -, 1, 127
John A. Wright, tenant, -, -, 2, 260
Bazel Wright, tenant, -, -, -, 75
James E. Dent, 100, 800, 3000, 75, 400

Wood County, West Virginia
1860 Agricultural Census

The University of North Carolina at Chapel Hill filmed the 1860 agricultural census for Wood County from originals at the West Virginia State Archives under a grant from the National Science Foundation in 1963.

Columns 1, 2, 3, 4, 5, and 13 represent the following information on the census:
1. Name of Owner, Agent or Manager of Farm
2. Acres of Improved Land
3. Acres of Unimproved Land
4. Cash Value of the Farm
5. Value of Farming Implements and Machinery
13. Value of Livestock

H. N. Crooks, 150, 150, 8000, 100, 400
Jas. Anderson, 15, 185, 1000, 10, -
Wm. Logsden, 70, 180, 3000, 250, 425
Jno. S. Anderson, 60, 140, 2000, 100, 350
Jno. W. Burdett, 35, 22, 1000, 25, 500
Elihue Burdett, 9, 18, 350, 20, 150
Luke Anderson, 30, 20, 1200, 20, 125
Samuel Williamson, 90, 250, 5000, 95, 200
Wm. Hardwick, 23, 17, 600, 8, 90
James Parks, 20, 50, 600, 3, 33
Wm. Lane, 30, 75, 1200, 15, 175
Mary Lane, 50, 100, 2000, 25, 100
Mrs. Barrows, 5, 45, 300, 3, 10
Rebecca White, 100, 240, 3000, 150, 400
John Smith, 33, 33, 600, 20, 200
Barns Smith, 25, 75, 600, 50, 150
Hyland Smith, 40, 50, 700, 10, 100
James Smith, 50, 190, 2400, 50, 385
Joseph Grim, 7, 125, -, 1, 30
John Boso, 70, 250, 3000, 50, 220
Wm. Mills Sr., 45, 45, 700, 30, 116
Michael Boso, 60, 20, 1000, 60, 530
Caleb Bailey, 20, 59, 600, 16, 110

Alfred Bowman, 70, 130, 1400, 60, 200
Albert Cosgrove, 60, 10, 1500, 50, 400
Wm. B. Leep, 50, 129, 1500, 50, 275
Samuel Leep, 8, 42, 200, 10, 25
David Kirkland, 20, 80, 500, 20, 124
Henry Sheets, 50, 200, 1500, 100, 325
Thos. A. Leep, 30, 70, 600, 125, 300
Lorenzo Leep, 30, 70, 600, 75, 200
Harrison Buckly, 200, 800, 5000, 150, 653
John Fleck (Fleek), 75, 125, 1500, 200, 387
Isaac Cotton, 50, 90, 1200, 100, 380
Cintha Beber, 30, 20, 350, 5, 75
K. N. McKinzie, 80, 270, 3000, 15, 187
Elijah Sheets, 25, 38, 380, 10, 128
John G. Biel, 20, 52, 570, 6, 50
Cyrus L. Anderson, 50, 325, 3000, 45, 295
Silas Hall, 40, 70, 880, 75, 265
Henry Cosgrove, 20, 30, 500, 100, 200
John Dewey, 36, 36, 600, 25, 27
Hiram P. Dewey, 15, 15, 300, 10, 87
Martina Flinn, 60, 23, 830, 25, 225

John Flinn, 140, 1160, 8000, 200, 1425

James M. Tracewell, 40, 610, 5000, 30, 135

Tabythas Smith, 15, 35, 400, 25, 174

Sol. W. Buffington, 50, 450, 3500, 95, 298

Gabreal Leep, 25, 60, 2300, 10, 140

Jas. Meddock, 40, 60, 800, 175, 141

Isaac Smith, 40, 60, 700, 30, 250

Lewis Provo, 30, 25, 350, 25, 85

Levinia Freeland, 40, 160, 500, 25, 159

Solomon Braham, 56, 244, 2000, 80, 270

Wm. Allen, 12, 63, 200, 10, 50

Cumming Smith, 20, 30, 200, 10, 50

Saml. T. Riel, 75, 258, 1700, 80, 322

Martin L. T. Burch, 40, 140, 900, 10, 85

Geo. Mills, 50, 50, 500, 25, 286

Michael Archer, 20, 37, 285, 10, 79

Wm. A. Buchanan, 60, 140, 1000, 60, 242

K. Atkinson, 15, 85, 500, 6, 35

Harvey Burch, 15, 241, 1000, 20, 113

Jackson Flinn, 45, 205, 1200, 115, 378

Peter Sutton, 10, 98, 690, 20, 114

Geo. Carter, 15, 85, 400, -, 95

Henry Justus, 8, 57, 120, 5, 30

Metilda Davis, 40, 250, 1200, 20, 158

James Low, 75, 440, 5000, 200, 404

Nathan Low, -, -, -, 10, 75

David Bowers, 30, 130, 800, 10, 338

John P. Low, Listed elsewhere, 10, 30

M. S. Anderson, 90, 510, 4000, 125, 890

Adam Bauman, 25, 35, 300, 50, 196

Geo. Bauman, -, -, -, -, 100

J. B. Roberts, 20, 30, 500, 50, 195

E. Anderson, 2, 20, 200, 5, 103

James Carson, 15, 85, 300, 10, 75

Moses Dunn, 4, 101, 309, 25, 136

Andrew Gates, -, -, -, -, 75

Wm. Pool, 30, 70, 400, 10, 128

John Price, 30, 70, 500, 10, 70

Charles Saunders, 10, 77, 400, 20, 80

Henry Pool, 30, 270, 1000, 40, 122

Levi Stephens, 70, 130, 1000, 60, 300

Asberry Montgomery, 15, 85, 450, 20, 160

Branard E. Coe, 80, 500, 3000, 20, 350

Wm. Beard, 60, 140, 800, 20, 283

Francis Roberts, 30, 24, 200, 5, 42

Wm. Brown, 25, 75, 400, 25, 200

James Burton, 28, 83, 600, 15, 105

John McIntyre, 40, 82, 600, 10, 117

E. Buchannon, 40, 110, 450, 15, 146

Geo. Lutes, 30, 70, 800, 25, 181

Tim Waters, 14, 36, 200, 3, 57

A. Buchanan, 30, 20, 300, 15, 40

Michael White, 20, 30, 300, 6, 100

Wm. Tribbit, 20, 52, 300, 10, 84

Joshua Buchanan, 25, 95, 40, 5, 38

John Lott, 50, 50, 800, 25, 285

Aaron Smith, 25, 75, 500, 15, 260

David Smith, 40, 88, 900, 20, 308

Wm. Smith, 45, 55, 900, 40, 300

Geo. W. Lott, 40, 50, 800, 10, 253

David Eaton, 60, 40, 800, 50, 360

Amos Eaton, 30, 152, 800, 100, 294

Benj. Amos, 60, 460, 3000, 50, 280

Robinson Ross, 20, 100, 300, 5, 34

Benj. Flerrence, 30, 340, 370, 25, 280

John Terrel, 35, 90, 450, 30, 17

George Flerrence, 20, 80, 500, 10, 53

Jacob Deem, 40, 236, 500, 10, 53

H. B. Deem, 60, 780, 3360, 75, 317

John Sheets, 40, 310, 1400, 10, 224

Aquilla Elliot, 6, 137, 715, -, -

Jacob H. Brown, 3, 72, 275, 10, 80

Benj. Aulman, 50, 350, 2000, 5, 55

Thomas Johnson, 60, 77, 400, 100, 134

Milton C. McLain, 25, 175, 3200, 50, 140

John N. Mathas, 15, 26, 235, 17, 74

Michael Lucas, 80, 94, 2000, 25, 220

John R. Leachman, 75, 100, 2000, 20, 146

H. S. Mitchell, 68, 109, 1700, 20, 323

John Henry, 50, 89, 1200, 20, 226

S. Coret, 50, 86, 1000, 85, 330

Jo. Smith, 20, 30, 450, 8, 50

S. Gilpin, 30, 137, 1200, 10, 28

J. E. Leary, 8, 122, 390, 23, 180

John Leary, 20, 80, 300, 50, 16

John Kellard, 15, 28, 129, -, 22

Wm. Lower, 35, 65, 400, 20, 136

Adam Hefler, 2, 28, 300, -, 65

Leander Lower, 17, 33, 250, -, 6

John Houser, 25, 125, 450, 10, 275

A. Givenn, 5, 70, 300, 3, 90

Geo. Wigal, 250, 400, 6800, 125, 662

Danl. Grogan, 75, 75, 1500, 75, 173

A. Grogan, 3, 57, 250, 10, 125

Elizabeth Palmer, 80, 82, 1500, 50, 263

Jacob Marty, 12, 88, 300, 5, 55

Benedict Burgy, 27, 73, 600, 80, 150

Jas. M. Dana, 20, 113, 600, 10, 100

Richard Phillips, 40, 60, 660, 20, 93

John Cannon, 50, 100, 1500, 20, 130

Samuel Emerick, 250, 390, 10000, 150, 890

Edmund Emerick, listed above, 15, 100

Hez. Emerick, listed above, 10, 120

Thomas Lower Sr., 100, 230, 1500, 150, 325

John Lower Sr., 70, 124, 1600, 20, 206

Stephen Johnson, -, -, -, -, 125

Mar__ Lower, listed above, -, 5, 60

George Lower, 30, 70, 500, 15, 150

Adam Henkle, 30, 36, 636, 20, 168

Harrison Henry, 35, 39, 800, 20, 200

Wm. Given, 35, 60, 70, 10, 160

A. Ingold, 30, 120, 800, 10, 100

Alex. Cole, 30, 90, 600, 10, 160

Wm. Hostotler, 5, 45, 550, 5, 80

John Lower Jr., 30, 74, 520, 10, 145

Elizabeth Flerrence, 30, 46, 330, 10, 108

Stephen Cason, 40, 460, 800, 40, 119

Adam Snider, 35, 65, 800, 3, 15

Danl. Snider, listed above, 26, 134

Hiram Snider, 25, 19, 220, 5, 140

Saphronia Quick, -, -, -, -, 85

Solomon Shears, 20, 113, 665, 5, 60

Anson Bloomer, 200, 127, 2000, 15, 235

Michael Dake, 12, 50, 250, -, 90

George Quick, 60, 193, 750, 10, 217

Philip Wile, 23, 73, 500, 5, 150

Justus Hanes, 25, 205, 800, 45, 232

David Lott, 25, 75, 800, 45, 232

Jas. J. Wigal, -, -, -, -, 180

Abram Wigal, 115, 385, 4000, 75, 700

Thos. Canary, 13, 37, 300, 5, 95

Fielding Phillips, 50, 78, 640, 15, 190

Peter Sellers, 40, 180, 1100, 18, 205

John M. Lowry, 30, 23, 250, 10, 100

Daniel Hanes, 30, 70, 600, 10, 145

A. Bleavens, 20, 79, 1000, 40, 139

J. D. Toomy, 30, 118, 700, 10, 190

John Gaslow (Gaston), 40, 83, 800, 50, 230

O. Logsden, 6, 94, 500, 3, 23

Abner Canary, 40, 104, 850, 5, 188

Elisha Dodson, 40, 85, 1000, 15, 218

Elias Rush, 40, 60, 800, 15, 243

Isaac Hostotler Sr., 25, 99, 600, 3, 200

Joseph Hale, 25, 75, 600, 4, 30

Philip Wigal, 100, 187, 2800, 100, 481

Jacob Houser, 30, 115, 900, 5, 208

Isaac Beber, 37, 41, 475, 6, 129

Saml. Ralston, 8, 60, 400, 4, 15

Saml. Swink, 16, 34, 300, 6, 20

Lorenzo Lafflin, 20, 69, 445, 10, 113

Samuel David, 20, 58, 400, 15, 98
Allen Robison, 25, 75, 500, 3, 100
Matthew Wuzar, 50, 87, 1500, 20, 158
J. Marlow, 60, 340, 4000, 80, 206
Roman Wuzar, 40, 45, 850, 75, 222
Elizabeth Wigal, 50, 180, 2000, 15, 241
B. Beckwith, 80, 120, 2000, 140, 435
S. G. Reeder, 12, 13, 215, 25, 55
S. Lightner, 25, 91, 800, 25, 148
E. Woodyard, 100, 190, 3000, 100, 254
E. Binner (Rinner), -, -, -, -, 100
Sarah Woodyard, 60, 190, 3000, 50, 374
Elijah Muncy, 45, 55, 1200, 40, 292
J. A. Ruble, -, -, -, -, 15
Geo. Fawkes, 20, 140, 300, 25, 77
George Cox, 14, 36, 300, 25, 77
D. S. Cook, 45, 55, 1000, 30, 154
Lewis Cramer, 35, 65, 400, 16, 75
Jacob Woodyard, 125, 129, 2500, 100, 500
J. Hostotler 2nd, 35, 145, 1440, 6, 100
J. S. Kesterson, 20, 80, 1000, -, 30
Hezekiah Davis, 40, 210, 1500, 10, 155
J___ Muncy, 35, 95, 1300, 10, 150
Robt. Collins, 50, 50, 1000, 75, 75
A. Anthony, 35, 15, 350, 45, 217
Danl. Stone, 85, 30, 3200, 100, 414
Eli Plankerton, 25, 79, 1000, 15, 110
Jos. H. Haddox, 18, 32, 500, 10, 95
Elizabeth Hoit, 100, 150, 5000, 125, 213
Wm. J. Cor (Cox), 40, 116, 1500, 10, 69.
Jno. Moyer, 23, 57, 800, 20, 104
James Treadway, 26, 31, 570, 100, 142
Pery Piggott, 10, 13, 184, 20, 129
Wm. Piggott, 5, 27, 200, 10, 40
James Thompson, 100, 244, 3000, 25, 271

Henry Hardman, 35, 95, 1300, 20, 257
A. Badgley, 35, 65, 1200, 55, 82
Wm. N. Davis, 20, 105, 700, 70, 150
Wm. Marvel, 60, 180, 2000, 50, 219
John Cooper, 110, 108, 3000, 75, 515
John Hill, 150, 200, 4000, 150, 1158
A. Devaughan, 16, 34, 1000, 25, 34
Philip Hardman, -, -, -, -, 24
Jas. Hardman, 20, 110, 1300, -, 100
P. N. Reeder, 25, 125, 1200, 20, 163
J. Sams, 25, 35, 400, 10, 100
Danl. Lee, 30, 120, 500, 50, 90
Wm. Cook, 40, 110, 1400, 30, 325
Stephen Lee, -, -, -, 20, 130
P. Ruble, 40, 60, 1000, 20, 15
D. Lee, 15, 289, 1600, 25, 130
Sarah Sams, 30, 44, 600, 20, 200
Barbara Sams, 40, 60, 1000, 50, 160
R. Black, 50, 125, 875, 75, 317
T. H. Reeder, 40, 34, 700, 70, 197
John Dye, 40, 57, 500, 15, 46
A. Givin, 25, 75, 500, 5, 94
John Givin, 30, 47, 350, 10, 75
Thos. Givin, 15, 57, 400, 10, 150
Saml. Givin, 50, 50, 500, 60, 271
Jas. Givin, 30, 67, 450, 15, 168
Stephen Givin, -, -, -, 20, 49
Wm. Black, 25, 75, 800, 40, 114
John Kinkade, 30, 225, 720, 50, 289
Jonnathan Sams, 30, 34, 500, 10, 98
J. Johnston, 80, 195, 1000, 20, 80
Geo. Stephens, 40, 10, 1000, 6, 158
Henrietta Pool, 40, 10, 1000, 10, 189
Richard Reeder, 100, 70, 3000, 100, 174
Sarah Stephens, 50, 50, 1000, 80, 211
Jas. M. Leech, 60, 140, 3000, 50, 394
Jas. Millrose, 40, 60, 1500, 50, 187
Wm. Leech, 15, 55, 400, 15, 117
Wm. W. Johnston, 60, 55, 600, 60, 293

Calvin Treadway, 24, 16, 800, 20, 133

Jas. Marshall, 30, 70, 500, 25, 170

A. D Reeder, 100, 300, 4000, 100, 636

Jas. Millrose, 80, 600, 5000, 50, 457

David McLain, -, -, -, 10, 115

John Millrose, -, -, -, 50, 150

Saml. Butcher, 25, 25, 500, 10, 40

Mark A. Millrose, 60, 1165, 3000, 45, 429

Thomas Graham, 50, 600, 2000, 25, 329

Danl. Henthorn, 30, 95, 400, 20, 314

Hannah Millrose, 30, 77, 750, 20, 100

Isaiah Lee, 25, 100, 1000, 100, 80

John Stephens, 60, 870, 6000, 100, 350

Jerrod Stephens, 6, 154, 800, 10, 145

Atwell Ruble, 70, 136, 558, 10, 137

Osias Stephens, 60, 140, 1000, 80, 260

Thos. J. Stephens, 18, 82, 700, 20, 122

Jacob Deem 4th, 20, 162, 910, 15, 90

Enoch Rector, 75, 10, 1000, 20, 282

D. H. Compton, 40, 95, 1800, 15, 391

Elizabeth Stephens, 80, 80, 2000, 15, 200

Thos. Stephens, 80, 20, 1600, 40, 308

Wm. Dawkins, 60, 170, 2000, 15, 269

Henry Cooper, 150, 86, 3000, 75, 511

James Barnett, 49, 12, 1000, 20, 235

John A. Page, 15, 103, 1080, 20, 223

Lewis Page, 80, 80, 1200, 50, 260

Lemuel Cooper, 100, 30, 3000, 50, 233

George Page, 100, 50, 2000, 100, 427

Robert Page, 40, 60, 1000, 15, 207

Saml. Pool, 20, 80, 500, 10, 191

Johnua Butcher, 40, 54, 500, 150,1 65

Hiram Deem, 150, 94, 2200, 50, 247

Henry Page, 70, 30, 2000, 20, 243

Henry Page, 75, 325, 4000, 50, 293

Wm. P. Deem, 7, 103, 550, 5, 120

Peter Deem, 70, 103, 2000, 50, 228

Miranda Deem, 30, 38, 340, 20, 200

E. P. Dye, 75, 200, 1500, 40, 317

Wm. Deem, 100, 1900, 6000, 50, 637

Thos. B. Hopkins, 20, 30, 200, 10, 90

Wm. R. Hopkins, 20, 30, 200, 10, 20

Joshua Exline (Epline), 8, 142, 350, -, -

Joshua R. Hopkins, 10, 90, 300, 5, 80

Jas. Anderson, 20, 130, 300, 10, 200

James Golden, 60, 740, 6000, 50, 265

Jonathan Steel, 60, 189, 2000, 20, 262

Alexr. Deem, 50, 320, 1100, 25, 45

Washington Berry, 50, 174, 1400, 30, 160

Thomas Berry, 50, 175, 1400, 100,157

Bazzle Wilson, 50, 100, 1300, 15, 182

Alfred Anderson, 20, 50, 1000, 14, 100

George Boice, 45, 255, 800, 30, 242

Michael Baker, 20, 380, 3000, 11, 158

Geo. Campbell, 60, 100, 1000, 20, 190

Joseph Weaver, 80, 100, 3000, 100, 227

Upton B. Johnston, 35, 162, 1000, 20, 200

Jonithan Deem, 45, 85, 1000, 10, 219

Jas. Bosner (Basner, Bames), -, -, -, -, 100

Ed. S. Butcher, 200, 400, 6000, 150, 658

Minedob Moor, 20, 82, 400, 10, 115

John T. Taylor, -, -, -, -, 75

H. C. Price, 3, 77, 240, 5, 165

Kawzada Curry, 22, 20, 600, -, 77

Adison Butcher, 210, 77, 5000, 100, 489

Geo. Barnett, 200, 100, 4000, 100, 933

Nathan Hutchison, 400, 100, 8000, 125, 1352

Willis Leech, 160, 160, 4000, 100, 341

Thos. Creel, -, -, -, 35

Jack Dawkins, 125, 260, 4000, 50, 495

John Cor, 50, 15, 500, 20, 190

Chas. Price, 80, 9, 1000, 100, 358

Richard Graham, 70, 52, 900, 50, 375

Vincent Dye, 30, 105, 300, 10, 109

John Page, 150, 100, 2500, 100, 578

John Barnett, 200, 241, 3682, 100, 526

John W. Tracewell, 60, 40, 2000, 25, 281

Benj. Cooper, 100, 105, 3000, 60, 380

Thos. Dawkins, 90, 110, 1600, 50, 350

Jas. Cooper, 100, 100, 2500, 200, 655

Wm. H. Taylor, 130, 140, 4200, 100, 510

David Pador, 60, 60, 4000, 25, 887

Jas. C. Athey, 60, 31, 2000, 50, 246

Zachariah Mann, 70, 115, 1885, 50, 186

Danl. C. Kincheloe, 80, 45, 1300, 75, 320

Peter Riddle, 65, 29, 2300, 50, 287

Solomon Linhart, 100, 100, 2350, 50, 360

Littleton Hall, 50, 46, 1000, 75, 213

Jacob Deem, 70, 22, 500, 20, 143

James Jackson, 45, 27, 700, 10, 142

Palsor Ruble, 60, 108, 1600, 20, 260

Jacob Ruble, 60, 40, 1000, 15, 100

Jacob Beynard, 141, 31, 4000, 30, 194

Adam Laughlin, 100, 50, 1500, 33, 713

James Blake, 50, 50, 3250, 130, 561

John Dulin, 50, 250, 1800, 15, 175

Wm. E. Posey, 60, 53, 1000, 40, 252

John Posey, listed above, 20, 146

Henson Lowe, 20, 45, 300, 10, 42

Nathan P. Williams, 30, 30, 240, 15, 153

N. G. Sole, 30, 26 280, 10, 136

Washington Sole, 30, 26, 280, 10, 247

Metilda Kisterson, 30, 34, 300, 6, 145

Jas. Armstrong, 100, 50, 1200, 100, 355

Otho Henry, 100, 160, 2000, 75, 245

John Satow, 30, 41, 600, 20, 207

Margaret Dashboyer, 5, 2, 200, 3, 70

Otto Shutz, 30, 70, 800, 10, 146

Saml. Manuel, 30, 145, 1200, 25, 271

Theodore Gebel, 4, 18, 150, 3, 60

Abraham Pennybacker, 30, 70, 800, 25, 237

H. Pennybacker, 40, 70, 900, 20, 110

J. Wigal, 60, 112, 1720, 50, 414

Desire Bridgley, 4, 1, 250, -, 73

Wm. B. Pennybacker, 40, 60, 1200, 15, 130

Jas. L. Daugherty, 40, 160, 1000, 6, 43

E. D. Dayton, 40, 260, 1200, 15, 237

Michael Carene, 15, 25, 200, 9, 91

Jonithan Coleman, 30, 70, 700, 5, 200

Robt. Hall, 16, 51, 800, 10, 125

Geo. W. Hall, 51, 149, 2000, 10, 280

Rebecca Hamelton, 2, 174, 575, -, 40

F. Kizer, 50, 83, 1200, 100, 315

F. Marlow, 100, 200, 3500, 75, 217
John Black, 18, 22, 200, -, 60
Jas. Wyal, 30, 120, 900, 10, 75
John W. Young, 51, 149, 2000, 10, 280
Lane A. Beckwith, 150, 350, 6000, 100, 870
Jane Beckwith, 80, 220, 3000, -, 55
Paton Cavenaugh, 45, 105, 900, 8, 125
E. Heydensicht, 16, 44, 600, 10, 69
E. Grogan, 41, 60, 1200, 6, 200
Hugo Heydensicht, 63, 69, 3000, 100, 335
Nancy Moor, 30, 110, 900, 4, 90
Jas. Doyle, 20, 45, 400, 15, 179
Wm. Spencer, 25, 50, 450, 8, 198
J___ Spencer, -, -, -, -, 75
Geo. Bird, 65, 211, 1380, 20, 391
R. H. Reeder, 40, 138, 880, 28, 274
David Sams, 40, 54, 500, 10, 276
Wm. Waggoner, 50, 80, 1200, 100, 232
Samuel Bartlett, 22, 44, 500, 10, 199
Adam Metheny, 40, 60, 800, 15, 155
John Matheny, -, -, -, -, 90
Wm. Lowe Jr., 16, 9, 250, 10, 275
J. B. Beckwith, 250, 227, 19000, 150, 850
Thos. M. Davis, 25, 75, 800, 6, 208
John Metter, 14, -, 800, -, 90
Willis Kesterson, 18, 26, 600, 5, 60
Truman Smith, 2, -, 500, -, -
P. Selby, 3, -, 400, -, -
Wm. Bartlett, 7, 18, 178, -, 15
T. F. H. Slevogt (Hoogt), 4, -, 900, -, 46
G. Robins, 6, -, 600, -, 17
J. S. Barnes, 1, -, 600, -, 95
Sol. Huff, 4, -, -, -, 40
Fred. Pail, 13, 7, 300, 9, 132
Jas. Romine, 30, 70, 1000, 20, 243
Silas Robins, 29, 21, 1000, 80, 223
B___ Pahl, 3, -, 300, -, 34
Alex. Woodyard, 22, 125, 1200, 5, 135

Henry Pahl, 3, -, 300, -, 34
Jonithan Robins, 200, 200, 4000, 1000, 513
Wm. Smitherman, 20, 127, 1176, 20, 112
Thos. Smitherman, -, -, -, 10, 190
Saml. Smitherman, 50, 100, 1200, 15, 152
Wm. A. Haslip, 25, 13, 380, 10, 170
Jedediah Ford, 50, 230, 2300, 100, 226
Hrace Cook, 100, 33, 2500, 70, 330
Wm. Bridges, 66, 10, 730, 100, 131
David Bridges, -, -, -, -, 75
Thos. Romine, 75, 5, 800, 40, 229
Saml. Romine, 100, 104, 2000, 100, 280
T. S. Dewey, 25, 75, 1200, 12, 127
Thos. Maddox, 120, 30, 1200, 50, 436
Wm. Killingsworth, -, -, -, -, 120
L. Q. Leavitt, -, -, -, -, 89
Benj. Robinson, 150, 165, 3500, 80, 580
Jas. Daggs, 3, -, 500, -, 55
F. Tavner, 66, 200, 4000, 100, 217
Thos. Tavner, 150, 200, 9000, 120, 520
Saml. Givin, -, -, -, -, 140
Calif Barrett, 100, 40, 2000, 100, 436
Jas. Rightmire, 140, 100, 4500, 80, 434
John C. Roberts, 200, 170, 6000, 100, 285
Wm. Reynalds, 150, 95, 3000, 100, 412
Wm. Parker, 300, 175, 4750, 80, 205
Wm. Nicely, 25, 75, 600, 24, 428
Henry Paw, 60, 40, 1000, 50, 329
H. N. Phillips, -, -, -, -, 30
Jas. Kesterson, -, -, -, -, 100
T. J. Howl, 20, 80, 400, 9, 115
J. D. Phelps, 12, 41, 265, 6, 30
Walker Mayhue, 140, 60, 3000, 125, 490

Michael Doyle, 54, 6, 1000, 75, 148

John Huested, 200, 60, 9000, 80, 1375

Jas. Smith, 50, 40, 1500, 15, 240

Elizabeth E. Chadock, 20, 20, 400, 3, 18

John Lutler (Sutler), 30, 16, 1500, 100, 234

R. C. Dodson, 12, 51, 720, -, 60

A. Dodson, 12, 51, 720, -, 60

A. Johnson, 5, -, 50, -, 124

Joseph Wilkins, 25, 75, 800, -, 32

Benj. Butcher, 75, 305, 2000, 26, 139

Benj. Edelin, 100, 127, 4000, 175, 643

Wm. Harwood, 125, 205, 10000, 314, 877

Isaac Fostner, 150, 50, 10000, 131, 677

Sarah Neale, 175, 25, 10000, 100, 795

Benj. Tracewell, 130, 70, 9000, 50, 620

Francis Lewis, 400, 200, 20000, 200, 1037

Mary McDugle, listed elsewhere, 436

Franklin Keen, 200, 47, 12300, 50, 515

Wm. Coffer, 155, 152, 5000, 79, 296

J. A. Baily, listed elsewhere, 1120

John McDugle, -, -, -, -, 60

Benj. Walker, 160, 110, 8000, 200, 515

Lemsy L. Grey, 15, 10, 300, 10, 120

Wm. Gray, 100, 150, 1900, 75, 260

Jas. L. Grey, 15, 11, 300, 15, 104

G. W. Wherry, -, -, -, -, 20

John Harwood, -, -, -, -, 151

Edwin Degenes, 50, -, 2000, 15, 268

Fred. Odenwahn, 18, 18, 673, 10, 36

Dominick Rockey, 16, 9, 700, 9, 108

Henry & Lewis Buckmier (Muchmier), 150, 75, 9000, 500, 958

Emil Meldahl, 150, -, 5000, 70, 1120

Jas. Dugan, 7, 18, 100, 2, 20

Asa Pras (Ras), 12, 78, 600, 40, 214

C. Byres, 5, 30, 400, 3, 100

Sylvester Ras, 20, 40, 500, 10, 190

Wm. R. Loads, 24, 11, 350, 10, 100

J. Lathrop, 20, 10, 500, 10, 120

H. Beadle, 15, 61, 732, 5, 45

Jas. Leavitt, 25, 87, 1120, 12, 126

Joseph P. Leavitt, 100, 260, 3600, 50, 328

E. W. Curtis, 60, 125, 3500, 40, 224

Robt. Kinchelos, 220, 140, 15000, 100, 981

Volney Tryon, 25, 61, 1000, -, 125

Elizabeth Wells, 103, 357, 10000, 370, 948

Geo. Mayberry, 300, 450, 20000, 400, 2195

Wm. Munchmire, 38, 16, 2000, 75, 475

Geo. R. Ruth, 2, -, 150, -, 22

J. M. Harris, 20, 170, 2000, 50, 222

Wm. N. Harris, 20, 170, 2500, 10, 80

Saml. Dewey, 12, 104, 1000, 20, 100

Wm. White, 42, 48, 900, 15, 114

Lucy Swindler, 40, 156, 1800, 10, 232

Wash. Wigal, 15, 850, 700, 5, 100

Henry Swindler, 80, 170, 4500, 50, 480

Benj. Brockheart, 60, 40, 2000, 20, 155

J. H. Leachan, 20, 50, 700, 15, 143

John Small, 12, 28, 400, 10, 93

Henry J. Hall, 13, 2, 700, 5, 43

Henry Hall, 40, 73, 2000, 25, 277

Jno. Darnbarger, 12, 55, 500, 15, 119

Jno. C. Harris, 30, 171, 1200, 58, 286

Jas. Johnston, 40, 67, 670, 75, 242

G. W. Buckly, 30, 120, 1000, 30, 175

Jno. Barton, 18, 82, 450, 5, 105

Smith Kindrew, 25, 25, 300, 6, 32

Geo. Flinn, 8, 142, 450, 6, 89

Nancy Harwood, 290, 445, 18000, 100, 560

J. A. Harwood, -, -, -, 60, 576

Henry Fields, -, -, -, 69, 230

Robt. Flinn, 40, 96, 1300, 12, 212

J. E. Williams, 100, 200, 3000, 40, 165

O. L. Bradford, 10, -, 1000, 10, 165

J. J. Jackson Jr., 59, -, 1500, 100, 805

George Neale, 150, -, 14000, 150, 1475

Geo. O. Pratt, 90, 93, 7000, 100, 296

Ben. Toothman, 70, 400, 7000, 140, 600

Jas. M. Davis, 40, 10, 1200, 12, 140

Hannah Samuels, 50, 150, 2000, 25, 254

A. S. Anderson, 2, -, 300, 3, 90

Eli McPherson, 40, 150, 800, 20, 255

G. W. Brooks, 75, 10, 3000, 20, 355

Plasting Bartlett, 14, 16, 400, 10, 65

Ugene Collett, 30, 119, 2000, 15, 75

Thos. H. Bartlett, 50, 117, 3000, 40, 548

Mary Bartlett, 40, 35, 1000, 5, 184

AnnVaughan, 25, 11, 500, -, 60

John Mount, 175, 104, 3500, 100, 293

Nesther Kinchelos (Kincheloe), 130, 160, 5650, 100, 559

Moses McGriger, 100, 51, 1500, 20, 180

B. W. Creel, 300, 100, 6400, 300, 761

B. H. Foley Jr., 100, 20, 2000, 30, 250

R. B. Kincheloe, 110, 40, 3000, 100, 337

Robt. Morrison, 100, 160, 2600, 20, 560

Z. Hickman, 30, 72, 1000, 10, 170

Isaac Cokely, 40, 70, 1110, 12, 172

Irvin Vaughan, 20, 80, 1000, 19, 300

Saml. W. Harris, 20, 23, 500, 16, 175

Webb Butcher, 150, 90, 3000, 28, 421

Saml. Warnick, 12, -, 240, -, 220

T. C. Byrd, 100, 300, 3000, 70, 316

Wm. Buckner, 12, 69, 1000, 10, 98

Alexr. Buckner, 65, 105, 2500, 30, 312

Robt. Buckner, 50, 30, 1500, 43, 466

Chas. Bebee, 66, 96, 1700, 50, 458

E. D. Stagg, 75, 79, 1540, 40, 381

Richd. Howard, 25, 25, 400, 15, 161

Nicholas Crouser, listed elsewhere, 50, 309

Gus Lazeur, -, -, -, 50, 275

John Bebbr, 65, 25, 1800, 75, 388

Tumer Boulware, 60, 40, 2500, 50, 288

Jeptha W. Bebee, 90, 84, 2500, 40, 364

John Hessmaman, 250, 150, 6000, 100, 1369

Thos. B. Steed (Stud), 75, 125, 600, 60, 105

Geo. W. Byrd, listed elsewhere, 75

Joshua Riley, 30, 220, 1200, 30, 100

Jas. Modisette, 15, 45, 500, -, 36

Geo. S. Riley, 11, 122, 250, 6, 45

Wm. P. Pickering, 45, 120, 900, 50, 294

A. C. Copelin, 12, 170, 600, -, 85

Orren Lower, 35, 83, 600, 20, 165

John Seaton, 40, 78, 600, 100, 220

John Vaughan, 30, 74, 500, 6, 76

David H. Riley, 30, 31, 400, 6, 80

John McManus, 30, 20, 600, 100, 196

Robt. H. Curry, 4, 86, 360, -, 18

Reuben Devaughan, 20, 110, 600, 6, 210

Thos. Devaughan, 30, 230, 1800, 85, 383

Isaac Littleton, 30, 108, 1000, 15, 120

Blackburn Wilson, 25, 125, 800, 16, 140

Catharine Hornback, 20,13,250, -, 75

John R. Croony (Curry), 25, 50, 400, 10, 100

Wm. Hickman, 70, 70, 2500, 16, 131

__. Terry, 30, 110, 444, 6, 181

Wm. Devaughan Jr., 40, 60, 1000, 30, 249

D. D. Davidson, 50, 27, 1200, 20, 341

G. S. Randall, 18, 50, 600, 12, 156

Wm. Hall, listed elsewhere, 20, 253

Wm. Devaughan Sr., 40, 30, 350, 20, 162

Bushrod Hall, 30, 40, 450, 10, 179

John Clay, listed elsewhere, 114

Fred Dancher, 15, 100, 600, 12, 133

E. R. Leech, 40, 20, 460, 13, 144

Benj. Hornback, 15, 55, 500, 6, 75

E. D. McGuire, 50, 256, 2448, 100, 578

Jacob Howl, 10, 90, 300, 2, 185

M. Holman, 30, 7, 1000, 15, 123

R. N. Smith, 45, 60, 840, 8, 143

Ezekiel Mount, 80, 120, 2000, 40, 231

John Dixon, 60, 350, 3000, 23, 275

Robt. Blacklin, 20, 124, 2000, 6, 145

Coonrod Gaines, 20, 30, 300, 20, 135

Jos. E. Gaines, 30, 45, 450, 27, 108

A. J. Burger, 140, 160, 2900, 20, 463

Elias Thornton, 50, 196, 2500, 40, 197

Wm. Thrash, 25, 150, 1000, 10, 194

Jerh. Reynolds, 30, 82, 500, 16, 194

Wm. Riley, 25, 25, 250, 9, 112

Nelson Bones, 34, 100, 700, 16, 190

John L. Harrow, 18, 24, 200, 3, 100

Isaac Harrow, 30, 60, 400, 25, 120

J. W. Bohard, 60, 99, 1150, 55, 351

John Founds, 5, 95, 400, 6, 80

Jacob Marshall, 50, 170, 1326, 50, 281

Mathias Howl, 30, 40, 600, 17, 213

Danl. Davis, 20, 54, 375, 10, 100

Zana H. Jones, 30, 500, 3000, 25, 140

Jos. Marshall, 25, 97, 615, 10, 158

Henry Ewing, 40, 284, 1600, 23, 116

Michael Mussetter, 30, 63, 600, 19, 122

Jacob Cork, 220, 205, 3000, 100, 1707

David J. Gabbert, 40, 160, 900, 10, 280

Err South, 100, 200, 2000, 25, 561

Wm. Marshall, 50, 200, 1700, 35, 265

Susan Rhoads, 30, 33, 400, 10, 25

G. B. Mount, 12, 10, 340, 10, 96

Wm. C. Womly, 100, 280, 3000, 10, 180

E. Whitlatch, 50, 148, 1200, 15, 225

Saml. Alton, 100, 300, 2000, 27, 300

Sampson Piersall, 75, 115, 1900, 25, 120

Mountjoy King, 15, 35, 350, 10, 80

Chas. King, 10, 42, 275, 6, 37

John A. Harris, 10, 50, 500, 8, 103

Columbus Mann, 50, 112, 900, 18, 140

Wm. Barrott, 45, 100, 1000, 25, 1442

Richard Johnson, 60, 140, 2000, 40, 286

Henry Lower, 70, 125, 1700, 50, 344

John Mann Jr., 75, 125, 1800, 100, 460

Geo. Harris, 60, 290, 3000, 85, 344

John Bennett, 65, 65, 1300, 50, 203

John R. Barrott, 25, 60, 850, 5, 123

Jacob Hiteshue, 110, 140, 2250, 100, 354

E. B. Vaughan, 25, 75, 1000, 15, 109

Jerry Stagg, 8, 86, 384, -, 80

Christian Grant, 20, 134, 500, 50, 159

J. M. Farsons, 18, 132, 950, 50, 192

Henry Farsons, 40, 94, 900, 25, 130

J. Piersall, 25, 160, 800, -, -

Robt. Grant, 75, 182, 200, 40, 360

G. W. Allman, 50, 210, 2100, 48, 312

Henry Dill, 12, 80, 450, 39, 60

Henry Musser, 25, 55, 320, 10, 116

John Farsons, 40, 108, 1200, 90, 235
F. Murphy, 10, 50, 600, 5, 280
David Linhart, 30, 218, 1000, 20, 160
Jno. S. Bicker, 40, 160, 800, 20, 325
David Finley, 40, 160, 600, 10, 117
Robt. Kincheloe, 40, 138, 1000, 115, 306
David Shears, 15, 155, 750, 9, 45
John Watkins, 70, 30, 1000, 20, 224
John Dixon, 50, 50, 600, 12, 200
Wm. Spencer, 40, 110, 300, 25, 346
Thos. Ward, 60, 33, 1400, 60, 160
Benj. Wining, 25, 84, 500, 10, 100
Saml. Kibbler, 70, 142, 2120, 50, 345
Andrew Murchy (Murdy), 100, 115, 1400, 65, 320
A. Kantinckes, 40, 65, 700, 12, 165
W. W. Wade, 75, 25, 2000, 10, 210
H. P. Dils, 550, 100, 13000, 200, 1978
Wm. Smith, 100, 80, 3800, 40, 266
Henry Dills 1st, 60, 50, 2200, 50, 247
Michael Cox, 40, 10, 1100, 27, 229
Jas. Stagg, 70, 32, 1200, 19, 165
Jno. Sutherland, 65, 16, 1100, 23, 258
Paul Cook, 100, 60, 320, 100, 858
E. P. Dils, 100, 200, 2300, 50, 315
Jos. Lyons 2nd, 90, 275, 3000, 100, 319
David Johnston, 12, 129, 1400, 40, 217
Catharine Johnston, 50, 350, 2500, 48, 280
Laurence Kincheloe, 75, 110, 3000, 50, 425
Josh Y. Smith, 30, 37, 1000, 20, 150
Benj. Chancelor, 25, 35, 600, 15, 164
H. M. Price, 30, 30, 900, 75, 235
H. H. Harper, 40, 60, 800, 40, 267
Jas. Atkinson, 80, 120, 1600, 100, 314
Uz Hoy, 100, 200, 900, 25, 270
Saml. Lutes, 65, 65, 1600, 40, 290

E. S. Rice, 30, 70, 800, 43, 218
Isaac Piper, 50, 100, 1200, 40, 246
Henry L. Stephens, listed elsewhere, 111
E. Johnston, 40, 140, 1100, 25, 185
Wm. Baily, 70, 142, 1300, 40, 325
Middleton Davis, 140, 224, 2190, 46, 484
Jas. Davis, -, 42, 168, -, 169
Isaac Yates, 45, 37, 700, 16, 280
Metilda King, 40, 37, 700, -, -
A. B. Nye, listed elsewhere, 185
F. T. Hendershot, 50, 67, 800, 39, 332
Danl. Grabel, 30, 84, 600, 24, 344
F. Little, 8, 49, 342, 100, 95
John W. Henry, 50, 37, 700, 15, 142
G. C. Griffin, 60, 48, 2300, 70, 287
Wm. Wood, 30, 64, 800, 23, 118
B. P. Ingram, 35, 220, 1200, 40, 203
Leml. Sinclair, 40, 91, 900, 25, 212
Henry Dye, 50, 94, 1000, 15, 182
James Dye, 25, 19, 260, 14, 200
Warner Smith, 35, 40, 395, 9, 267
Jos. Stephens, 50, 135, 925, 20, 318
E. W. Whitlatch, 40, 30, 350, 10, 95
John Bowley, 25, 65, 600, 12, 100
Jas. Whitlatch, 50, 50, 900, 23, 173
Saml. Boxmore (Bosemore), 60, 140, 1500, 4, 365
Jacob Lemly, 30, 100, 520, 15, 184
Wm. Stoops, 75, 375, 1700, 25, 269
Jas. Stoops, 12, 353, 500, 10, 151
Saml. Minor, 30, 69, 300, 15, 133
A. Headly, 55, 205, 1040, 19, 278
Jos. Bradford, 10, 40, 125, 6, 121
Jacob Garrison, 30, 70, 600, 10, 145
Ezra Johnston, 35, 25, 360, 12, 157
Josiah Hutchison, 30, 30, 500, 15, 167
Elmer Wilkinson, 50, 285, 1100, 20, 231
Jas. Stephenson, 30, 152, 900, 12, 252
Johnson Garrison, 15, 85, 450, 10, 18

Thos. Whitlatch, 35, 165, 600, -, 21
Benj. Prince, 4, -, 50, -, -
Jabez Swiger, 20, 37, 400, 5, 141
David Whitlatch, 19, 61, 350, 10, 100
Jonithan Johnston, 100, 250, 2500, 50, 300
Ellis Johnston, listed above, 18, 354
Elza Johnston, listed above, 8, 112
Thos. Sinclair, 20, 30, 500, 10, 157
Fredk. Sinclair, 25, 25, 500, 10, 120
John A. King, 40, 100, 1400, 19, 262
Lewis Ogden, 70, 239, 1854, 50, 331
Phillip Linch, 10, 140, 700, 10, 169
Lewis M. Ingram, 20, 80, 1100, 15, 174
Adam Darling, 170, 130, 4500, 50, 517
John Sharp, 100, 170, 2000, 60, 395
Loyd Riddle, 150, 120, 9000, 75, 290
Rufus Ralston, 242, 1076, 12328, 375, 1268
Nathan Ralston, 200, 840, 7627, 50, 516
Wm. McKinny, 59, 44, 3000, 175, 429
Robt. Burke, 15, 88, 600, 10, 180
Wm. Hunter, 70, 100, 1330, 60, 350
A. Griffin, 60, 67, 1270, 60, 275
Elihue Davis, 65, 110, 1750, 15, 98
Wm. P. Davis, 25, 50, 750, 10, 224
Wm. Barrons, 40, 60, 1000, 25, 193
R. Pollock, 40, 40, 800, 56, 185
Jos. Barrons, 30, 70, 900, 10, 58
David McKibben, 40, 60, 1000, 54, 241
John McKibben, 35, 33, 1000, 125, 250
S. M. Peterson, 80, 125, 2050, 75, 242
Wm. E. Stevenson, 100, 84, 2750, 130, 246
Abraham F. Ingram, 120, 1410, 5300, 100, 483
Alexr. Gault, 50, 39, 1000, 70, 971

R. S. Corbit, 165, 208, 7500, 125, 971
J. A. Smith, 45, 28, 2900, 100, 2500
C. A. Williamson, 60, 58, 3450, 60, 160
Sarah Sisson, 70, 78, 4400, -, -
Jas. A. Berry, 50, 114, 3200, 50, 360
Geo. Compton, 100, 500, 6000, 100, 363
Jas. Davis, 25, 25, 1000, 4, 193
Henry Janes, 70, 175, 2800, 30, 259
R. Kinnaird, 140, 139, 8370, 100, 770
John Hazlerigg, 143, 196, 10000, 50, 520
G. Ogden, 30, 20, 1500, 25, 220
J. W. Snodegrass, 62, 21, 4000, 225, 180
Danl. Hoover, 20, -, 400, 20, 106
John Athey, 30, 445, 1000, 16, 163
A. Kinnaird, 40, 145, 1000, 16, 163
D. J. Hazlerigg, 15, 15, 3000, 45, 153
Wm. Page, 40, 34, 600, 40, 282
Lewis Roe, 75, 163, 1200, 30, 280
Thos. Hoover, 80, 68, 1480, 50, 213
Thos. Bunch, 75, 75, 1500, 30, 295
J. M. Reed, 45, 355, 2000, 43, 300
W. M. Ashby, 15, 15, 333, 5, 90
Thos. Ashby, 16, 39, 500, 10, 114
G. S. Henry, 18, 12, 300, 15, 1441
John Henry, 50, 150, 2000, 25, 275
H. E. Hultz, 18, 22, 675, 10, 100
Ben Athey, 40, 60, 900, 16, 188
Danl. Halder, 50, 100, 1200, 25, 120
Danl. Christian, 40, 125, 1000, 50, 189
Saul Chitester, 25, 75, 600, 16, 227
Cyrus Pugh, 75, 125, 1900, 75, 217
Walter Locker, 30, 11, 410, 30, 174
T. W. Locker, 80, 120, 2000, 45, 312
Jacob Bogas, 50, 50, 1000, 13, 218
Sardes B. Athey, 45, 55, 1000, 25, 122
Zela Athey, 80, 56, 1400, 50, 638
Mary A. Uhl, 30, 30, 630, 3, 47

Robt. Pugh, 75, 120, 2500, 115, 309

Enos Pugh, 45, 35, 2000, 28, 242

W. Hoover, 45, 43, 1000, 10, 190

Jesse Pugh, 63, 7, 2100, 30, 155

Rodolph Uhl, 25, -, 750, 40, 209

Chas. D. Uhl, 108, 120, 3600, 50, 314

G. W. Henderson, 275, 350, 30000, 325, 1290

Jas. Tomlinson, 370, 630, 50000, 240, 1196

Morgan Henry, 60, 100, 1600, 25, 275

Jas. T. Uhl, 40, 10, 1500, 23, 237

Geo. Uhl, 50, 100, 2500, 5, 42

Danl. Brookover, -, -, -, -, 150

John T. Johnson, 140, 160, 10000, 125, 649

Rachael Keller, 45, 79, 2500, 60, 377

Wm. Johnson Jr., 200, 73, 8500, 250, 1260

David Uhl, 40, 20, 1800, 15, 270

Abraham Johnson, 95, 61, 4500, 15, 288

Sardes Cole, 50, 100, 3000, 25, 161

A. N. Cole, listed elsewhere, 50, 339

O. Hiet, 50, 60, 5500, 45, 226

Jas. Hiet, 160, 20, 9000, 130, 220

J. S. Stone, 115, 58, 8600, 20, 495

J. Stapleton, 125, 27, 10000, 200, 1000

Delila Grear, 200, 250, 16000, 300, 650

Wm. Spencer, 130, 700, 20000, 300, 768

J. W. West, 80, 48, 10000, 200, 546

T. A. Cook, 70, -, 6000, 25, 80

Edward Johnson, 50, -, 3750, 100, 141

Oliver Owen, 20, 111, 786, 25, 226

E. T. Bartlett, 98, 2, 6500, 30, 132

R. H. Lord, 50, 53, 5000, 185, 415

E. C. Neal, 130, 128, 12000, 25, 115

Alfred Neale, 60, 100, 5000, 125, 420

Thos. Meeks, 50, 75, 5000, 75, 535

Wm. H. Neale, 100, 50, 10500, 250, 288

A. Fuller, 75, 43, 9000, 525, 610

T. J. Cook, 130, 30, 10200, 350, 884

Elizabeth Neale, 54, 6, 3500, 10, 20

Chas. Rust, listed elsewhere, 175

Mantine Brookes, 13, 22, 200, 8,100

Jas. Rion, 25, 5, 400, 25, 146

Dilla Shaw, 21, 4, 300, 15, 206

Goodall Dare, 35, 25, 900, 20, 230

Ezekiel Dye, 14, 50, 1300, 25, 360

G. W. Gell, 75, 47, 1800, 15, 165

Thos. Harkins, 100, 200, 2400, 100, 378

Joshua Johnson, 141, 410, 5500, 200, 459

Jordan Hall, 53, 39, 1800, 100, 220

John D. Simms, 80, 7, 1500, 120, 318

John Anderson, 60, 90, 1200, 75, 248

Jas. T. Fulton, 75, 25, 2000, 50, 220

Saml. Ogden, 21, 11, 640, 10, 151

Wm. Cooley, 50, 75, 1250, 25, 92

L. Mott, 58, 12, 1400, 63, 295

E. Taylor, 110, 40, 6160, 100, 397

H. Woodyard, 75, 75, 3000, 10, 119

B. H. Foley, 220, 80, 15000, 250, 980

Mason Foley, 140, 80, 4500, 95, 595

Osena Woodyard, 30, -, 900, 10, 154

B. T. Beeson, 185, 43, 15750, 200, 495

B. Cook, 100, 9, 14700, 150, 363

Wm. Wigal, 60, 140, 2000, 50, 314

John Kincheloe, 150, 75, 1000, 150, 720

Wm. Smith, 60, 23, 2000, 40, 180

Danl. Gardway, 37, 38, 1500, 73, 150

James Harwood, 100, 10, 8000, 125, 340

S. H. Williamson, 45, 73, 1800, 40, 300

Wyoming County, West Virginia
1860 Agricultural Census

The University of North Carolina at Chapel Hill filmed the 1860 agricultural census for Wyoming County from originals at the West Virginia State Archives under a grant from the National Science Foundation in 1963.

Columns 1, 2, 3, 4, 5, and 13 represent the following information on the census:
1. Name of Owner, Agent or Manager of Farm
2. Acres of Improved Land
3. Acres of Unimproved Land
4. Cash Value of the Farm
5. Value of Farming Implements and Machinery
13. Value of Livestock

This county had a large number of tenants.

Wm. Roach, 45, 155, 2500, 15, 140
James B. Cook, 50, 135, 1500, 10, 300
Leroy B. Chambers, 53, 10, 1500, 200, 1500
Thos. G. Cook, 50, 175, 500, 10, 284
Alderson Cook, 75, 200, 900, 12, 180
Wm. D. Cook, 50, 50, 1200, 10, 620
Robinson Cook, 30, 100, 600, 15, 300
James H. Cook (tenant), 8, 73, 200, 5, 100
Thomas Toler, 40, 272, 2233, 35, 578
John Brown, 15, 100, 250, 3, 230
Thomas Bailey, 55, 319, 2170, 50, 220
Isaac Bailey (tenant), 6, 67, 469, 4, 114
Wm. Lusk (tenant), 16, 84, 900, 8, 169
Archabald Bailey, 15, 85, 300, 2, 45
Louden Bailey (tenant), 25, -, 125, 25, 83
Morison Cook (tenant), 30, -, 300, 5, 179

David Cook, 97, 200, 1770, 150, 2680
Wm. P. Brown, 20, 30, 300, 4, 103
Floyd P. Cook (tenant), 22, -, 450, 20, 981
Mathias Brown, 15, 6, 200, 2, 40
Isaac M. Toler, 30, 175, 375, 5, 114
James M. Elswick (tenant), 8, 257, 175, -, 15
John Cook Esq., 100, 1500, 4700, 300, 564
Wm. Cook (tenant), 40, 52, 800, 10, 200
Sanders Mullins (tenant), 20, 10, 350, 60, 21
George Webb, 60, 320, 600, 3, 200
Nathaniel Blankinship, 40, 293, 263, 4, 100
Aden Thompson, 80, 302, 1500, 50, 850
Jacob Cook, 100, 240, 4500, 120, 1835
Isaac Cook, 125, 600, 7500, 90, 625
David Toler (tenant), 10, -, 150, 5, 121
Wm. H. Cook, 150, 1550, 7500, 100, 514
George Milum, 25, 82, 250, 10, 110

Wm. Walker, 25, 50, 500, 15, 300
Geo. Stewart, 40, 148, 600, 95, 225
Charles Stewart, 175, 589, 3500, 100, 505
Matterson Bailey, 50, 350, 400, 10, 158
Green M. Cook, 75, 400, 2400, 10, 319
Robert Stewart, 75, 500, 1200, 20, 200
Hubbard Meadows, 20, 55, 300, 2, 90
Mitchell Cook, 30, 700, 600, 12, 144
Madison Elis (tenant), 25, 100, 400, 6, 60
Henry Clay (tenant), 45, 125, 600, 2, 15
Thos. W. Sizemore (tenant), 8, 167, 250, -, 32
Thos. Laxton, 45, 1000, 1250, 10, 190
John L. Cook, 20, 100, 738, 5, 88
Wm. N. Cook, 30, 401, 600, 10, 215
John H. Stewart, 35, 300, 500, 7, 70
Thos. M. Cook, 20, 300, 500, 15, 87
Sylvester Cook, 40, 496, 1300, 6, 27
John W. Browning, 10, 30, 80, -, 23
John Mullins Sr., 30, 36, 600, 12, 403
John Mullins Jr., 20, 25, 400, 10, 215
Jacob Brinegar, 22, 704, 910, 8, 212
Colvin Sizemore, 22, 116, 300, 5, 35
David Goode, 40, 229, 600, 10, 205
Henry _. Clay, 50, 100, 300, 100, 186
Joseph Mitchell, 25, 125, 200, 10, 243
John S. Mullins, 85, 700, 200, 25, 600
Charles Workman (tenant), 20, 150, 500, 8, 175
Shadrich Green (tenant), 16, 25, 400, -, -
Charles Mitchell, 35, 100, 400, 8, 90
Hiram Lambert, 60, 400, 1800, 6, 482

Arthur Buchanan, 25, 60, 500, 10, 116
Joshua Green, 20, 93, 200, 8, 158
Alfred Green, 20, 125, 203, 3, 93
Barnabas Evans, 20, 123, 400, 9, 188
Simeon Green, 16, 84, 300, 7, 125
E. Sizemore (tenant), 20, 80, 250, 3, 53
G. P. Sizemore, 15, 1753, 876, 10, 120
Thos. Godfry, 30, 355, 700, 25, 297
John Stadly, 30, 34, 400, 6, 160
Patton White (tenant), 20, 80, 100, 8, 117
Martin Bailey (manager), 12, 88, 100, 5, 26
Wm. Bailey, 30, 110, 400, 20, 230
James O. Bailey, 20, 30, 300, 45, 189
Henry F. Lusk, 25, 300, 800, 30, 230
Floyd Lusk, 35, 290, 800, 5, 231
Gordon C. Lusk, 25, 85, 392, 8, 139
John Bolcher(Belcher), 20, 176, 250, 3, 52
Alexander Lane, 15, 90, 600, 8, 82
John Garison, 30, 70, 500, 40, 752
P. G. W. Rineheart, 100, 100, 1200, 25, 254
Marshal Mullins, 15, 110, 500, 20, 125
John Howerton, 40, 500, 1000, 200, 162
Rhuben Howerton, 15, 35, 500, 10, 135
Powetan McKiney, 45, 60, 700, 10, 274
Wm. Coleman, 43, 150, 900, 7, 165
Reuben Roach, 25, 175, 600, 20, 184
Samuel P. McKiney, 24, 100, 400, 6, 172
Lusk Graham (tenant), 50, 130, 500, 1, 460
Jacob Akers, 25, 70, 100, 5, 100
George W. Solesbury, 30, 70, 500, 5, 321
James Graham, 18, 82, 300, 6, 111

M. V. Solesbury (tenant), 65, 535, 1000, 10, 403

Silas Hatcher, 72, 278, 1000, 40, 128

Robert Mills, 50, 607, 1000, 20, 904

James Wiley, 50, 180, 1300, 20, 384

Christian Walker, 55, 145, 1000, 30, 324

Chrisprianess Walker, 60, 1020, 5000, 26, 226

Nancy Walker (tenant), 18, 125, 700, 5, 31

Bird Lester, 50, 300, 1200, 77, 594

Acelis Fanning, 30, 35, 350, 10, 226

H. J. Garetron, 18, 282, 500, 4, 37

Hugh J. Fanning, 12, 88, 200, 5, 98

Elijah Meadows, 25, 75, 300, 3, 88

Wm. Meadows (tenant), 20, -, 100, 4, 115

Charles Walker, 80, 400, 1700, 100, 650

James Waddle, 17, 133, 500, 8, 96

Hugh M. Justice, 28, 70, 400, 20, 172

Philop Hedrick, 45, 350, 1500, 150, 450

Wm. Fink, 135, 430, 3000, 200, 680

Eliza G. Cole, 30, 245, 500, 3, 209

Pleasant Lilly, 40, 460, 500, 8, 348

Wm. E. Cole, 15, 585, 400, 2, 165

David R. Cox (tenant), 15, 200, 100, 4, 35

Jonas Bragg, 30, 168, 654, 8, 187

Mathus Maxwell, 100, 600, 2100, 150, 600

James N. Wood (tenant), 25, -, 125, 10, 120

James Cline, 25, 115, 150, 5, 125

Jarimiah Solesbury, 35, 107, 800, 15, 425

Daniel H. Agee, 25, 143, 500, 20, 292

Thos. H. Covey, 40, 200, 400, 6, 137

Green W. Bailey (tenant), 25, 375, 300, 9, 108

Jeremiah Covey, 40, 114, 900, 10, 190

John O'neal, 35, 565, 600, 10, 100

John Dunn (tenant), 18, 3, 141, 10, 295

D. B. Covey (tenant), 18, 300, 150, 40, 449

James W. Cook (tenant), 50, 100, 900, 10, 217

John Acord (tenant), 35, 100, 300, 15, 281

Valentine Meadows (tenant), 10, 90, 100, 1, 25

Burwell McComas, 15, 20, 500, 7, 35

Franklin Sizemore, 40, 173, 300, 15, 168

George F. Workman (tenant), 16, 40, 200, 5, 23

John Terry, 18, 18, 225, 12, 35

Wm. Evans, 35, 155, 450, 15, 355

M. F. Hill (tenant, 35, 400, 2000, 50, 30

Bartly Rose, 30, 690, 1500, 20, 474

Amos Shumate, 33, 30, 500, 10, 260

Wm. Short, 36, 125, 800, 10,187

Thos. M. Birchfield (tenant), 20, 20, 200, 8, 30

Madison Workman, 15, 100, 300, 20, 136

Hiram Toliver (tenant), 25, 75, 175, 15, 280

Wiley Phillops, 50, 130, 736, 20, 564

Wm. G. Phillops (tenant), 16, 119, 476, 10, 222

John Meadows, 28, 57, 500, 12, 169

R. M. Cook, 12, 63, 400, 6, 35

Austen Cooper, 12, 390, 600, 10, 79

Thos. Cook, 100, 196, 1500, 100, 415

Elett Cook, 10, 130, 1500, 18, 30

R. R. Roach, 12, 8, 60, 10, 219

Nemiak Allen (tenant), 22, 50, 400, 20, 145

Chloe Mandeville, 210, 9000, 3800, 50, 263

John Canterbury, 75, 175, 1300, 100, 350

John Duncan (tenant), 15, 20, 500, 5, 68

Wm. Stewart, 60, 22, 1000, 25, 220

Wm. Brooks, 50, 150, 600, 60, 355

Andrew Garnoe, 60, 281, 200, 50, 976

Jacob Walter (manager), 30, -, 300, 100, 244

Harison Cuzar (tenant), 18, -, 200, 5, 52

Zur Acord (tenant), 18, -, 180, 15, 181

John McCraw, 75, 175, 2500, 70, 945

Goodall Dier, 30, 70, 800, 10, 281

Nancy Acord, 30, 100, 400, 9, 200

G. F. McMillion (tenant), 18, -, 150, 6, 90

Peter G. Farmer, (tenant), 30, 1000, 500, 5, 215

Robert Acord (tenant), 25, 78, 500, 10, 150

John Allen Jr., 40, 95, 400, 15, 240

J. Q. Brooks (tenant), 16, 500, 300, 10, 70

Randolph Brooks (tenant), 30, 500, 500, 10, 180

John Allen Sr., 75, 175, 1500, 100, 700

Britten Allen, 40, 54, 1200, 20, 351

Josiah Cook, 28, 36, 600, 10, 262

Lavis Cook Esq., 75, 175, 600, 8, 237

Patk. Cook Esq., 70, 30, 1500, 150, 600

Henderson Baily, 30, 470, 600, 40, 449

Jacob Glandon, 40, 196, 600, 150, 130

Wm. Cornut, 15, 125, 350, 25, 114

Larkin Collins, 30, 350, 300, 5, 325

Henry Elis, 100, 1300, 2500, 150, 1444

Larkin Smith, 60, 930, 2000, 15, 663

Isaac Roberts, 25, 35, 200, 3, 217

Martin Lester, 20, 276, 500, 5, 416

Michael Cline, 75, 1425, 1500, 75, 1510

George Whitt, 25, 25, 200, 20, 69

Eli Whitt, 10, 20, 100, 7, 17

Harvey Bailey, 40, 315, 1000, 12, 1350

Floyd Morgan, 12, 88, 150, 5, 148

Eli Blankinship (tenant), 20, 100, 500, 7, 115

Jessee Davis, 20, 75, 300, 16, 217

Isaac R. Lester, 12, 100, 500, 3, 157

Pleasant Lester, 35, 365, 1200, 5, 151

Joseph Lester, 60, 500, 2240, 40, 317

Armer L. Godfry, 25, 75, 300, 10, 273

James Adams, 12, 88, 200, 3, 27

George Walker, 50, 300, 1000, 4, 690

Wm. Shields, 25, 150, 500, 5, 280

Lane Shannon, 75, 300, 1500, 110, 446

Joseph McDonald, 480, 2414, 13000, 115, 380

Wm. McDonald, 370, 3666, 16000, 200, 4015

E. McDonald (manager), 300, 1447, 13000, 200, 2910

David Osbourn (tenant), 10, 290, 500, 2, 35

James Morgan (tenant), 10, 65, 100, 5, 113

George Johnston, 60, 639, 5299, 60, 964

Lorenzo D. Walls, 14, 46, 200, 2, 210

Wm. Gremmett, 15, 30, 200, 5, 114

David Morgan, 75, 217, 2000, 50, 309

John Toler, 25, 175, 300, 20, 220

Charles Toler, 12, 263, 475, 10, 25

Thos. Morgan (tenant), 15, 15, 300, 5, 95

James Hatfield, 18, 57, 300, 2, 75

Joseph Walls, 50, 100, 600, 12, 274

Squire Toler, 38, 126, 600, 40,160

Alexander Varny, 35, 237,800, 10, 242

Valentine Hatfield, 40, 40, 700, 10, 504

Wm. Toler, 75, 527, 1000, 10, 261

Ballard Shannon, 44, 100, 600, 22, 45

Levi Gore, 24, 106, 800, 125, 656

James H. Shannon, 60, 450, 2000, 36, 697

Wilson Harvey, 28, 125, 600, 14, 388

Andrew Harvey, 18, 50, 200, 6, 250

Milton Morgan (tenant), 50, 35, 400, 25, 186

Ransom Harvey, 15, 60, 175, 5, 149

Mathew Raney, 35, 75, 600, 5, 265

Patterson Harvey, 8, 50, 200, 5, 85

James Shannon, 250, 1145, 3000, 200, 985

Abraham Belcher, 35, 60, 150, 6, 120

Bartly Belcher, 20, 20, 15, 15, 110

Haven Bane (manager), 115, 1000, 3000, 100, 1382

James Bailey, 70, 400, 1500, 125, 590

George Morgan, 45, 85, 1000, 8, 220

Andre Hatfield (tenant), 55, 381, 1000, 20, 155

Mary Blankinship (tenant), 25, 100, 300, 4, 100

David Bishop, 20, 130, 150, 20, 145

Henry F. Jackson, 11, 87, 300, 7, 163

Reece T. Lusk, 50, 300, 1000, 200, 365

Index

Bames, 118

Bandy, 104

Bane, 131

Banner, 108

Bannister, 2

Bar, 103

Barb, 73

Barber, 78-79, 82

Barbour, 7

Barbrug, 94

Barburg, 94

Bardine, 102

Bargerhoff, 69

Barker, 32, 34, 44, 60, 95, 99

Barkheimer, 55

Barkish, 80

Barnes, 48, 109, 120

Barnet, 6

Barnett, 90-91, 118-119

Barnhart, 81

Barr, 65

Barrackman, 74

Barrett, 120

Barrons, 125

Barrott, 123

Barrows, 114

Bartlett, 27, 34, 48-49, 77, 97-98, 120, 122, 126

Barton, 121

Bartrug, 94

Bartrum, 80, 83-84

Bashard, 50

Basner, 118

Bassel, 97

Bassell, 71, 74

Bastable, 66, 72

Bates, 106

Bathgate, 23

Batson, 44

Batten, 39, 81

Baughman, 67, 90-91

Bauman, 115

Baxter, 101

Baylis, 2

Beaden, 35

Beadle, 121

Bean, 73-74

Bear, 67

Beard, 115

Beasly, 67

Beaty, 58

Beauchamp, 109

Beauford, 7

Beauty, 107

Bebbr, 122

Bebee, 122

Beber, 114, 116

Becket, 96

Beckett, 4

Beckly, 9

Beckner, 78

Beckwith, 117, 120

Bedlman, 60

Bee, 32-34

Beeson, 126

Belamy, 86

Belcher, 128, 131

Beleard, 108

Bell, 15, 22, 32, 35, 94, 110, 112

Bender, 89

Beneditt, 42

Benee, 122

Benner, 110

Bennet, 46

Bennett, 12, 25, 66, 75, 77, 107-108, 123

Bennington, 32

Bent, 40

Berlin, 71

Berria, 66

Berry, 118, 125

Bessard, 100

Bessey, 35

Best, 5

Bett, 110

Beverage, 76

Beverlin, 110

Beynard, 119

Bias, 2, 4

Bice, 65

Bicker, 124

Bickerstaff, 36

Flesher, 4, 40-41, 58-59
Fleshman, 11
Fletcher, 44, 63
Fling, 35
Flinn, 28, 114-115, 121-122
Flint, 67
Floyd, 68, 94
Fluharty, 29
Fogg, 103
Foggy, 16
Folan, 100
Foley, 122, 126
Folks, 15
Folsburg, 18
Folsbury, 18
Ford, 2, 46-49, 120
Fordyce, 64
Fornash, 66, 72
Forth, 3
Foster, 2, 32, 68, 71-72, 74, 111
Fostner, 121
Fought, 106, 111-112
Foughty, 111-112
Founds, 112, 123
Fowler, 27, 90
Fox, 32-33, 60, 97
Frampton, 81
Francis, 9
Frank, 60
Frankenberger, 25
Franks, 102
Frasher, 83, 86
Frazer, 2-3
Frazure, 112
Fredericks, 1, 35
Freeland, 43, 60, 95, 115
Freeman, 1, 44-45, 91
French, 28
Frest, 48
Fretwell, 72, 75
Fricket, 107
Frieadly, 29
Friendly, 29
Friel, 76
Frinnell, 39
Fronsman, 68

Frum, 45, 62
Fry, 85
Frymier, 66
Fulerton, 81
Fulkerson, 100
Fulks, 104
Full, 106
Fuller, 86, 126
Fulton, 126
Fultz, 69, 74
Fulwider, 31
Furbee, 63, 99
Furbey, 100, 102
Furguson, 10
Furr, 24
Furrow, 10
Fury, 74
Gabbert, 123
Gadd, 89
Gadolee, 73
Gaheny, 110
Gain, 98
Gainer, 18, 100
Gaines, 123
Galaway, 59
Gallaspie, 3
Gandee, 40
Gandell, 37
Gannon, 4
Garden, 94
Gardener, 7
Gardner, 89
Gardway, 126
Garetron, 129
Garison, 128
Garlo, 101
Garman, 59
Garner, 6, 27-28, 91, 101-103
Garnett, 103
Garnoe, 130
Garnson, 111
Garrett, 78-79
Garrison, 33, 124
Garten, 12
Garter, 77
Gary, 47

Gaston, 116
Gates, 48-49, 108, 115
Gatlin, 47
Gault, 125
Gaws, 86
Gawthrop, 44-45, 48, 77
Gbourn, 35
Geary, 37
Gebel, 119
Gecman, 33
Gell, 126
George, 23, 63, 66, 73, 106
Gerley, 97
Gerry, 5
Gibson, 2, 37-38, 72
Gilbert, 48
Giles, 6, 46
Gilkison, 80, 82
Gilmer, 72
Gilmore, 53
Gilpin, 116
Gimms, 2
Girgin, 33
Girodend, 6
Given, 88-89, 91-92, 116
Givenn, 116
Givens, 43
Givin, 117, 120
Gladwell, 74-75
Glandon, 130
Glaslow, 116
Glenn, 45, 104
Glivens, 96
Glover, 34, 62, 94-95, 97
Godby, 12
Goddard, 100, 103
Godfry, 24, 128, 130
Godolee, 73
Goff, 8, 34-35, 39, 41-42, 46, 51, 53,
91
Goin, 33
Golden, 30, 118
Good, 37
Goodall, 12
Goode, 128
Gooden, 74

Goodin, 32
Goodrich, 98
Goodwin, 56
Goone, 7
Goove, 7
Gorbey, 98
Gorby, 104
Gorden, 19
Gore, 3, 12, 131
Gorney, 2
Gorrel, 58-59
Gorrell, 25
Gothing, 110
Gould, 69-71, 74
Gowen, 65
Gowers, 66
Grabel, 124
Graham, 10, 100, 118-119, 128
Granden, 57
Grass, 4-5, 11, 123
Gratehouse, 49
Graves, 63, 70-71
Gray, 7, 11, 45, 51, 109, 121
Grayson, 24
Grear, 126
Greathouse, 40-41, 43
Green, 37, 77, 128
Greene, 88, 90-91
Gregg, 28, 61, 63
Gregory, 67, 88-89, 91
Gremmett, 130
Grey, 121
Griffin, 27, 77, 88, 91, 124-125
Griffith, 3, 73
Grigsby, 89
Grim, 16, 60, 104, 106-107, 114
Grimes, 47, 67
Grimm, 72
Grims, 74
Grizzle, 83
Grogan, 116, 120
Grose, 103
Gross, 70
Groves, 74
Grubb, 72
Grump, 96, 98

Gulley, 66
Gum, 30, 67, 76, 91
Gumm, 33
Gump, 95
Gurley, 1-2
Gusman, 48
Guthrie, 1, 99, 101-102
Guyer, 77
Hacker, 82
Hadan, 41
Haddon, 81
Haddox, 23, 28, 32, 75, 95, 117
Haden, 33
Hadley, 23
Hafer, 101
Hagans, 32
Hague, 25
Haid, 42
Hail, 111
Hainor, 25
Halbert, 109
Halder, 125
Hale, 12, 116
Hall, 3, 22-23, 25-26, 29, 32-33, 35, 40-41, 76, 91, 96, 100-101, 103, 107, 111-114, 119, 121, 123, 126
Hally, 111
Halsted, 12
Hamack, 88
Hamelton, 119
Hamer, 14-15
Hamilton, 15, 20, 24, 61
Hamlin, 60
Hammack, 37, 40
Hammer, 24, 31-32
Hammon, 34, 42
Hammond, 60, 108
Hamner, 70, 76
Hamond, 98
Hampton, 87
Hamrick, 29, 88-89, 91
Hanaman, 109
Handlen, 102
Handley, 2, 5, 39
Handly, 82
Hanes, 61, 99, 103, 116

Haney, 80, 101-102
Hanger, 43
Hanna, 91
Hannah, 67, 69
Hanner, 110
Hannon, 109
Hansford, 53
Harbargar, 32
Hardbarger, 22
Harding, 19
Hardinger, 47
Hardman, 35, 39, 42, 117
Hardwick, 114
Hare, 15, 60
Hareflinger, 103
Harisford, 14
Harkins, 126
Harlace, 39
Harlas, 70
Harless, 40
Harman, 7-8, 42-43, 81-82
Harmer, 7
Harmmon, 40
Harold, 32, 38
Harper, 10, 16-17, 19, 39-40, 43, 52, 72, 76, 124
Harr, 45-46
Harris, 6, 18, 26, 28, 31-34, 72-73, 75, 77, 96, 106, 111-112, 121-123
Harrison, 5-6, 78
Harrow, 123
Harsh, 52
Hart, 14, 16, 18-19, 33, 102
Hartley, 6, 59, 101
Hartman, 66
Harvey, 5, 67, 131
Harwood, 121-122, 126
Hascue, 38
Hasker, 94
Haslep, 58
Haslip, 120
Hass, 46
Hatfield, 23, 129-131
Hatton, 82, 85-86
Hauger, 43
Haught, 62, 108, 110

146

Perkins, 42, 61
Perrill, 107
Perrin, 112
Perrine, 90-91
Perry, 75-76, 84, 86-87
Pervel, 110
Peters, 35, 46, 66, 68, 81
Peterson, 28, 125
Petit, 93
Pettigrew, 99
Petty, 107, 109-222
Pew, 23, 26
Phares, 16, 18-20
Pharis, 17-19, 101
Phelps, 120
Philips, 22, 25, 47, 60, 94
Phillips, 16-17, 43, 52-53, 69, 74-76, 116, 120
Phillops, 129
Phipps, 75
Pickens, 20
Pickergill, 105
Pickering, 111-112, 122
Pickersman, 111
Piercy, 19
Pierpoint, 32, 58-59
Piersall, 123
Pifer, 52, 72
Piggott, 117
Piles, 25, 80, 83-84, 98
Pinson, 85
Piper, 69, 124
Pipes, 58
Pitman, 10
Pitts, 62
Pittsford, 5
Pitzenbarger, 106
Plankerton, 117
Plum, 108
Plumly, 9, 12
Plybon, 78
Plymale, 82
Poe, 47-48, 96
Pogue, 6
Poindexter, 2
Point, 35

Pollock, 125
Polly, 4
Pomroy, 113
Pool, 29, 40, 115, 117-118
Poor, 2
Porter, 52, 57, 84-87
Posey, 119
Post, 66, 68-69
Postlewate, 97-100
Postleweight, 24
Potsy, 10
Potts, 15, 76, 103
Powell, 50, 101, 103
Powers, 1, 18
Pras, 121
Prather, 32-33
Pratt, 24-26, 63, 87, 100, 122
Presley, 42
Preston, 86
Pretty, 2
Pribble, 22, 30
Prible, 111-112
Price, 23, 26, 39-40, 68, 80, 87, 95, 104, 110, 115, 119, 124
Prichard, 58
Prickett, 58
Priddy, 6
Prince, 9-10, 34, 125
Pringle, 69, 71
Prirey, 18
Pritchard, 32-33
Pritt, 70, 76-77, 90
Probest, 104
Procter, 89
Propst, 24
Proudfoot, 18
Provo, 115
Prunty, 29-30, 34
Pugh, 89, 125-126
Pumphry, 71-72
Pursinger, 4, 7-8
Quarles, 3
Queen, 16, 65-68, 83-84
Quick, 16, 116
Qunlin, 12
Racer, 2

Weese, 13, 16-18, 20
Weese, 90-91
Welch, 26, 41
Wellman, 82-83, 86
Wells, 10, 24, 26, 28-28, 55-56, 59, 61, 63, 109, 121
Welsh, 2, 5, 72
Wentz, 69
West, 5, 40, 48-49, 94-96, 109-110, 126
Westfall, 17, 32, 34-35, 66-67, 69, 73-74
Weton, 55
Wetzel, 96
Whaly, 25
Wheeler, 4, 44, 46, 53, 64
Wherry, 121
White, 6, 14-17, 19-20, 38, 51-52, 63, 66-068, 71, 73, 101, 107, 114-115, 121, 128
Whitecotton, 17
Whited, 40
Whitehair, 49
Whitehell, 2
Whiteman, 22, 103-104
Whitescotton, 111
Whitison, 51
Whitlatch, 123-125
Whitt, 87, 130
Whittaker, 45
Whitten, 104
Whittington, 6
Whitzel, 39
Wiatt, 7
Wickham, 100
Wicks, 72
Wigal, 116-117, 119, 121, 126
Wigner, 22-24, 26, 29
Wilcox, 64, 81
Wile, 116
Wilex, 81
Wiley, 8, 83-85, 129
Wilfong, 14, 68, 73-74
Wilkins, 121
Wilkinson, 7, 80, 102, 113, 124
Willey, 94-95, 98

Williams, 6, 9-11, 17, 30-31, 47, 57, 60, 72, 81, 90, 97, 102-103, 112, 119, 122
Williamson, 25, 29, 31, 45-46, 55-56, 87, 114, 125-126
Wills, 11, 27
Wilmoth, 13, 18-19
Wilson, 15, 17, 25, 28-30, 32-33, 35, 39, 45, 47, 67, 75-76, 81-83, 85-87, 107-112, 118, 122
Wimer, 32
Winemiller, 69-71, 75
Winer, 108
Winerich, 22
Wingfield, 74
Wining, 124
Wirt, 110
Wise, 104
Wiseman, 29, 46, 105, 108
Wissons, 1
Witcher, 83
Withrow, 8
Wolf, 16, 34, 39
Womly, 123
Wood, 11, 14-15, 29, 47, 72-73, 124, 129
Woodbourn, 62
Woodburn, 58
Woods, 28-29, 89-91, 93, 98
Woodsides, 28
Woody, 1, 7
Woodyard, 48, 108-109, 117, 120, 126
Woofter, 33
Woolf, 52, 112
Woolford, 53
Woolwine, 19
Wooton, 85
Wootring, 52
Wordon, 5
Workman, 6, 11, 18, 59, 85, 101, 128-129
Wrick, 27
Wright, 6, 34, 39-40, 60, 63, 94, 101, 112
Wuntendon, 5

www.ingramcontent.com/pod-product-compliance
Lightning Source LLC
Chambersburg PA
CBHW080614270326
41928CB00016B/3049